This volume presents a unique synthesis of our understanding of modern cosmology. It is based on a series of highly successful lectures Professor Rees gave under the auspices of the Accademia Nazionale dei Lincei to review the progress of cosmology and its future prospects.

Observational cosmology has made remarkable advances in recent years and has brought into sharper focus a new set of fundamental questions that Professor Rees addresses in this book. Why is the universe expanding the way it is? What were the 'seeds' that caused galaxies, clusters and superclusters to grow? What is the nature of 'dark matter'?

Many key issues and current controversies are also considered, but throughout a clear distinction is maintained between aspects that now have a firm empirical basis, and those that remain speculative. Given the broad contribution Professor Rees has made to cosmology, this book will be welcomed both by researchers in the field and by beginning graduate students with a background in physics.

Perspectives in Astrophysical Cosmology

Lezioni Lincee
Sponsored by *Foundazione IBM Italia*
Editor: Luigi A. Radicati di Brozolo, Scuola Normale Superiore, Pisa

This series of books arises from lectures given under the auspices of the Accademia Nazionale dei Lincei and is sponsored by *Foundazione IBM Italia.*
The lectures, given by international authorities, will range on scientific topics from mathematics and physics through to biology and economics. The books are intended for a broad audience of graduate students and faculty members, and are meant to provide a '*mise au point*' for the subject with which they deal.
The symbol of the Accademia, the lynx, is noted for its sharp-sightedness; the volumes in this series will be penetrating studies of scientific topics of contemporary interest.

Already published

Chaotic Evolution and Strange Attractors: D. Ruelle
Introduction to Polymer Dynamics: P. de Gennes
The Geometry and Physics of Knots: M. Atiyah
Attractors for Semigroups and Evolution Equations: O. Ladyzhenskaya
Asymptotic Behaviour of Solutions of Evolutionary Equations: M. I. Vishik
Half a Century of Free Radical Chemistry: D. N. R. Barton in collaboration with S. I. Parekh
Bound Carbohydrates in Nature: L. Warren
Neural Activity and the Growth of the Brain: D. Purves

Perspectives in Astrophysical Cosmology

MARTIN REES

Institute of Astronomy, Cambridge

CAMBRIDGE
UNIVERSITY PRESS

Published by the Press Syndicate of the University of Cambridge
The Pitt Building, Trumpington Street, Cambridge CB2 1RP
40 West 20th Street, New York, NY 10011-4211, USA
10 Stamford Road, Oakleigh, Melbourne 3166, Australia

First published 1995

Printed in Great Britain at the University Press, Cambridge

A catalogue record for this book is available from the British Library

Library of Congress cataloguing in publication data

Rees, Martin J., 1942–
Perspectives in astrophysical cosmology / Martin Rees.
p. cm. – (Lezioni Lincee)
ISBN 0 521 47530 9. – ISBN 0 521 47561 9 (pbk.)
1. Cosmology. 2. Galaxies. 3. Nuclear astrophysics. I. Title. II. Series.
QB981.R37 1995
523.1 – dc20 94-24454 CIP

ISBN 0 521 47530 9 hardback
ISBN 0 521 47561 9 paperback

VN

Contents

Preface

This small book is based on a series of Lezioni Lincee presented in Milan. In the lectures I tried to outline, for an audience of physicists as well as astronomers, some aspects of current research at the interface between extragalactic astrophysics, cosmology, and particle physics: topics addressed include galaxy formation, the origin of structure, dark matter, the background radiation, etc.

The presentation was superficial, and glossed over many key points. This was primarily through my own inadequacy, but was to some extent inevitable in any attempt to cover such a range of issues in only six hours. The present written text re-orders (and in some places updates) the lectures. However, in deference to the tradition of brevity established by earlier publications in this series, I have not expanded the material beyond the level of detail that could actually be presented in the lectures. The list of references, for the same reason, is not fully comprehensive (though it includes a selection of 'further reading').

This is definitely not a 'textbook', and cannot serve as a self-contained primer. But I have tried to highlight what seem the most important results and ideas (though Chapters 4 and 5 are somewhat more eclectic, being focussed on some more specialised topics I was working on at the time). I hope that, as a whole, the lectures conveyed the essence of some recent developments and current debates, without too much distortion, and that this written version of them will provide specialists in other branches of physics, and students coming to the subject for the first time, with an accessible introduction and overview.

It is a pleasure to acknowledge the influence of many colleagues with whom I have collaborated or discussed cosmological topics. I am grateful to the Accademia Nazionale dei Lincei for inviting me to give the original lectures, and to Professor Ettore Fiorini for being such a supportive and hospitable host in Milan.

1

The cosmological framework

Introduction

Gravity, almost undetectable between laboratory-scale bodies, is the dominant force in astronomy and cosmology. The basic structures in our cosmic environment – stars, galaxies, and clusters of galaxies – all involve a balance between gravitational attraction and the disruptive effect of pressure or kinetic energy. Our entire observable universe may display a similar balance: the Hubble expansion is being slowed (and may perhaps eventually be braked to a halt) by the gravitational effect of its entire mass-energy.

The best-understood cosmic structures are the smaller ones: the individual stars. Stellar structures and life cycles can be predicted theoretically, and tested empirically by observing large populations of stars, of differing ages, in the Milky Way. The Milky Way, the disc galaxy to which the Sun belongs, can be envisaged as a kind of ecological system in which stars are continually being born and dying, their gaseous content being recycled and chemically enriched as the evolution proceeds.

Our own Galaxy is typical of the galaxies distributed through the universe, which are the most conspicuous features of the cosmic scene. Why should the universe be full of these remarkable aggregates of stars and gas, typically $\sim 10^5$ light-years across and containing around 10^{11} stars? We do not yet have compelling physical reasons for the characteristic properties of galaxies, as we do for stars.

One reason why galaxies are harder to understand than stars is that their formation impinges on *cosmology*. Individual stars form, evolve, and die more or less regardless of what the universe does – initial cosmic conditions have left no traces on the complex gas dynamics that goes on within each galaxy. But that is not true for galaxies, which may have emerged, at an epoch when the entire universe was denser and perhaps very different, from inhomogeneities that were imprinted on the universe in its earliest phases.

Large-scale structure: how homogeneous is the universe?

In the perspective of the cosmologist, even entire galaxies are little more than 'points of light' which indicate how the material content of the universe is distributed, and how it moves. Galaxies are clustered: some in small groups (like our own Local Group, of which the Milky Way and the Andromeda galaxy are the dominant members), others in big clusters with hundreds of members. Moreover the clusters themselves are grouped in filamentary or sheet-like superclusters. In recent years, there has been great progress in quantifying the distribution of galaxies over the sky, and also in mapping out the three-dimensional structure. The latter task has entailed determining redshifts and distances for thousands of galaxies.

Figure 1 shows the major groupings of galaxies within our local part of the universe, out to a distance of around 3×10^8 light-years. At still greater distances, galaxies are distributed more uniformly over the sky. There is no evidence that big density contrasts extend to larger scales. A region of the size shown in this figure is therefore probably large enough to provide a fair sample of the contents of the universe.

Our universe certainly does not have a simple fractal structure, with clusters of clusters of clusters *ad infinitum*. There is a definite upper limit to the scale on which large-amplitude density inhomogeneities are observed. The *largest* structures with $(\delta\rho/\rho) \gtrsim 1$ are about 1 per cent of the Hubble radius: the

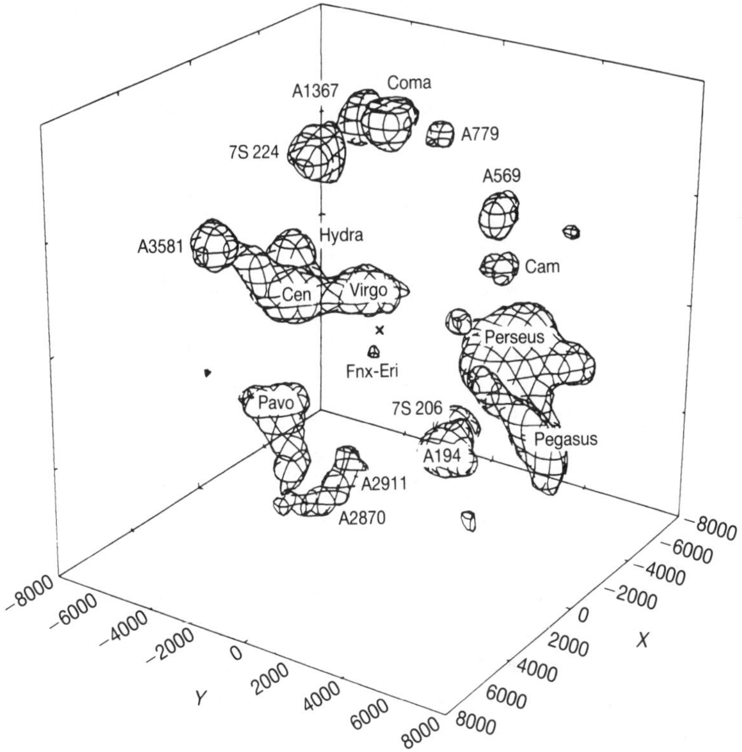

Figure 1 The most conspicuous clusters and superclusters within a cube, of dimensions around 3×10^8 light-years (10^8 pc), centred on our own Galaxy. There are also, of course, many galaxies more uniformly distributed in the space between these clusters. The linear dimensions of the region depicted here are about 2 per cent of the size of the part of the universe accessible to optical observations. This cube is probably large enough to provide a 'fair sample' of the contents of the universe: on larger scales the amplitude of inhomogeneities is much less than unity. (From Hudson, M. J. 1993, *Mon. Not. Roy. Astron. Soc.* **265**, 43 (Fig. 10).)

typical nonlinear scale is around 0.3 per cent. The typical metric fluctuations due to clusters and superclusters – defined in dimensionless form as the gravitational energy per unit mass arising from the associated density enhancement, in units of c^2 – have an amplitude of the order of $\psi = 10^{-5}$. The velocities relative to the Hubble flow induced by these structures are typically below $\psi^{1/2}c = 1000$ km s^{-1}. The mass-equivalent of the kinetic energy associated with these so-called 'peculiar motions' is therefore only 10^{-5} of the rest mass. This number $\psi = 10^{-5}$, a measure of the metric fluctuations in our universe, has an importance which will come up again later. Its smallness implies that the present local dynamics of cosmic inhomogeneities such as clusters and superclusters can be validly approximated by Newtonian gravity. Even more importantly, it justifies the relevance of the simple theoretical models for a homogeneous isotropic universe. These models date back to the 1920s.

The first relativistic models for a homogeneous expanding universe were found by Friedmann[1] before Hubble[2] discovered the recession of the nebulae. Hubble's work, which showed that the universe did not resemble Einstein's earlier static model, stimulated further studies of relativistic cosmology by Lemaitre, Tolman, and others. But the data were then – and remained for several decades – too sparse to indicate whether any of these idealised models fitted the real universe, still less to discriminate among them.

High-redshift objects

Hubble's work suggested that the galaxies would have been crowded together in the past, and emerged from some kind of 'beginning'. But he had no direct evidence for cosmic evolution: indeed the steady-state theory,[3] proposed in 1948 as a tenable alternative to the 'big bang', envisaged continuous creation of new matter and new galaxies, so that despite the expansion the overall cosmic scene never changed.

To discern any cosmic evolutionary trend, one must probe objects so far away that their light set out when the universe was significantly younger. This entails studying objects billions of light-years away with substantial redshifts. A programme to measure the cosmic deceleration was pursued from the 1950s onwards with the 200-inch Palomar telescope.[4] But the results were inconclusive, partly because normal galaxies are not luminous enough to be detectable at sufficiently large redshifts. It was Ryle and his colleagues from radio astronomy,[5] in the late 1950s, who found the first real evidence that the universe was indeed evolving. Radio telescopes could pick up emission from some unusual 'active' galaxies (which are now believed to be harbouring massive black holes in their centres) even when they were too far away to be seen with optical telescopes. One cannot determine the redshift or distance of such sources from radio measurements alone, but Ryle assumed that, statistically at least, the ones appearing faint were more distant than those appearing intense. He counted the numbers with various apparent intensities, and found that there were too many apparently faint ones – in other words, sources at large distances – compared with the number of brighter and closer ones. This was discomforting to the 'steady statesmen', but compatible with an evolving universe if galaxies were more prone to undergo violent outbursts in the remote past, when they were young. The subsequent discovery by optical astronomers of extreme 'active galactic nuclei' (quasars) at very large redshifts corroborated Ryle's conjectures, but these objects, and their evolution, are still too poorly understood to be used for determining the geometry of the universe.

By probing deep into space, astronomers can study parts of the universe whose light set out a long time ago. If we lived in a wildly inhomogeneous universe, there would be no reason why these remote regions (and the way they have evolved) should

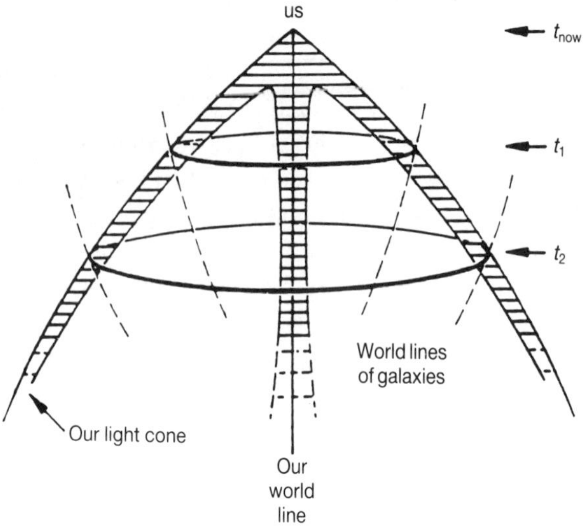

Figure 2 Schematic space-time diagram showing world line of our Galaxy and our past light cone. The only regions of space-time concerning which we have direct evidence are those shaded in the diagram, which lie either close to our own world line (inferences on the chemical and dynamical history of our Galaxy, 'geological' evidence, etc.) or along our past light cone (astronomical evidence). It is *only* because of the overall homogeneity that we can confidently assume any resemblance between the distant galaxies whose light is now reaching us and the early history of our Galaxy. In homogeneous universes we can define a natural time coordinate, such that all parts of the universe are similar on hypersurfaces corresponding to a given value of t.

bear any resemblance to our own locality. However, insofar as the universe we find ourself living in (or at least the part of it accessible to observation) is actually uniform and isotropic, its gross kinematics are describable by a single scale factor $R(t)$; all parts have evolved the same way and have the same history (see Figure 2). This simplicity gives us reason to believe that when we observe a region of the universe that lies (say) 3 billion light-years away, its gross features (the statistical properties of the galaxies, the nature of the clustering, etc.) resemble those that would have been displayed 3 billion years ago in our own locality (i.e. within the region depicted in Figure 1).

Astronomers have an advantage over geologists, in that they can directly observe the past. Optical or near-infrared images recorded on CCD solid-state detectors, 50 times more sensitive than photographic plates, reveal huge numbers of very faint galaxies – up to 100 per square arcminute – each of which is so far away that its light set out before our Solar System formed. These galaxies appear typically only 2 or 3 arcseconds across. To image them sharply requires the full resolution of the Hubble Space Telescope. Now that this instrument has been effectively repaired it should soon reveal whether these relatively young galaxies have the same morphology as those around us. What fraction of them have, for instance, already acquired discs?

Even more remote than these very faint galaxies are some of the quasars and other 'active galactic nuclei' (e.g. the intense radio sources). These are the hyper-active centres of a special class of galaxies, so bright that they outshine the stellar content of the galaxy by a factor of many thousands. The highest redshift so far[6] is $z = 4.89$ for PC 1247+3406, whose spectrum is shown in Figure 3. The wavelengths of the radiation from this quasar have been stretched, between emission and reception, by a factor $R_{now}/R_{em} = 1 + z = 5.89$; the Lyman-α 1216 Å line is therefore observed in the red part of the spectrum, at around 7200 Å. To estimate the relative age of the universe then and now, one needs to know the dynamics of the expansion, and in particular how much it has been decelerating. If there were no deceleration at all, the universe would have been 'younger' when the light set out by the same factor 5.89. However, according to the Friedmann models the expansion is decelerating. In the theoretically attractive Einstein–de Sitter cosmology

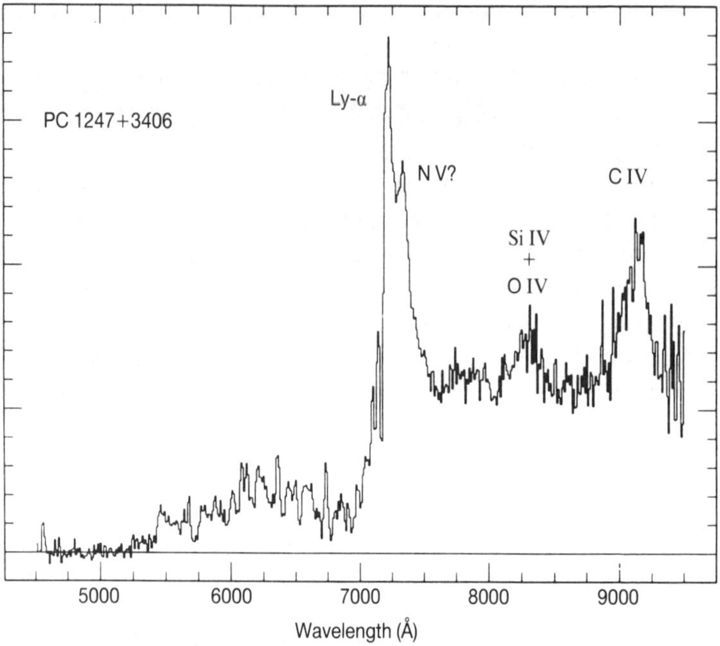

Figure 3 The spectrum of the quasar PC 1247+3406, with redshift $z = 4.89$. Light from this object set out towards us when the cosmic scale factor R was $(1 + z) = 5.89$ times smaller than it is today. According to the Einstein–de Sitter model, the universe would then have been only ~7 per cent of its present age. (From Schneider, D. P., Schmidt, M. & Gunn, J. E. 1991, *Astron. J.* **102**, 837.)

the scale factor of the universe grows as $R \propto t^{2/3}$. The light now reaching us from PC 1247+3406, according to that model, would have set out when the universe was younger by a factor $5.89^{1.5}$. Astronomers can therefore probe the last 90 per cent of cosmic history. The existence of these quasars tells us that by the time the universe was about 10^9 years old some galaxies (or at least their inner regions) had already formed, and runaway events in their centres had led to the extreme type of active nuclei that the quasar phenomenon represents.

The light from quasars offers an important probe for the intervening medium. Absorption lines blueward of Lyman-α in the spectrum indicate clouds of gas lying along the line of sight.[7] The absorption is probably caused by protogalaxies too faint to be seen by their direct emission (and where perhaps no stars have yet formed). The way this absorption depends on redshift offers important clues to how galaxy formation proceeded; this is discussed further in Chapter 5.

Pre-galactic history

But what about still earlier epochs, before any galaxies could have formed? Did everything really emerge from a dense (or even singular) 'beginning' ten or twenty billion years ago? The clinching evidence dates back to 1965, when Penzias and Wilson published their classic paper announcing 'excess antenna temperature at 4080 Mc/s'. Intergalactic space is not completely cold but has a temperature of about 3 K. This may not sound much, but it implies that there are about 4×10^8 photons per cubic metre – maybe a billion photons for every atom in the universe.

The discovery of the microwave background quickly led to a general acceptance of the so-called 'hot-big-bang' cosmology – a shift in the consensus among cosmologists as sudden and drastic as the shift of geophysical opinion in favour of continental drift that took place at about the same time. There seemed no plausible way of accounting for the microwave background radiation except on the hypothesis that it was a relic of an epoch when the entire universe was hot, dense, and opaque. Moreover, the high intrinsic isotropy of the radiation meant that the simple mathematical models were a better approximation to the real universe than the theorists who devised them in the 1920s and 1930s would have dared to hope. Subsequent measurements of this background, made with increasing precision at various wavelengths, have strengthened these conclusions. The radiation spectrum is now known, primarily through the magnificent results from Mather and his collaborators,[8] using the FIRAS (Far Infrared Absolute Spectrophotometer) experiment on the COBE (Cosmic Background Explorer) satellite, to deviate from a black body by less than 1 part in 10^4. The best-fitting temperature is 2.726 K. And measurements by several groups[9–12] show that the radiation is intrinsically isotropic to within a few parts in 10^5, but that there are apparent anisotropies, on angular scales from 1° up to 90°, at the 10^{-5} level (some quantitative implications of these are mentioned in Chapter 3).

In the dense early phases, the radiation would have been held in thermal equilibrium with the matter scattering repeatedly off free electrons whose density would have been high enough to make the universe very opaque. But when expansion had cooled the matter below 3000 K (when the cosmic scale factor R was $10^{-3}R_{\rm now}$) the primordial plasma would have recombined, leaving few free electrons. The 'fog' would then have lifted, the universe thereafter becoming transparent, and probably remaining so until the present (see p. 93). The experimentally detected microwave photons are direct messengers from an era when the universe was about a thousand times more compressed, and had expanded for about half a million years. But the photons are still around – they fill the universe and have nowhere else to go. An important 'cosmic number' is the photon-to-baryon ratio η^{-1}, which stays essentially constant during the cosmic expansion. It is because this ratio is large that many authors refer to the *hot* big bang.

The universe contains other important fossils of a cosmic era far earlier than (re)combination: the light elements such as D, ^{3}He, ^{4}He, and ^{7}Li. During the first *minute* of cosmic expansion, when temperatures were above 10^9 K, nuclear reactions would have synthesised these elements, in calculable proportions, from protons and neutrons. The baryon density in an expanding universe goes as $R^{-3} \propto T^3$, and would therefore have been 10^{27} times higher when $T = 3 \times 10^9$ K than it is today. But this is still not as high as the density of air! One does not need to worry about problems of dense matter. And the energies of the relevant nuclear reactions are < 1 MeV, and do not involve any large uncertain extrapolation from the experimental domain. Such calculations,[13] showing how the light-element abundances would depend on the present mean baryon density, the number of neutrino species, etc. were done in the 1960s. Although refinements have been introduced,[14] nothing essential has changed on the theoretical front over the last 20 years.

Stellar nucleogenesis, supernova explosions, and recycling into new stars, the theory of which was formulated in the 1950s,[15] seem well able to account for 'heavy' elements such as carbon, oxygen, and iron.[16] But the high and relatively uniform proportion of helium always posed a problem. It was therefore gratifying, and neatly complementary, that helium was the one element that would be created prolifically in a 'big bang'. In the 1970s, the astrophysical problems of accounting for deuterium (whose abundance is reduced during stellar recycling) were properly appreciated, and this isotope is also believed to be a cosmological fossil.[17]

Only more recently have astronomers been able to determine the light-element abundances in old stars, gaseous nebulae, etc. precisely enough to make a worthwhile comparison with the 'big-bang' predictions. In particular, the helium abundance is now pinned down with a precision approaching 1 per cent. What is remarkable is that, as Figure 4 shows, the light-element abundances all appear concordant with the predictions of 'big-bang nucleosynthesis', provided that the baryon density is in the range 0.1 to 0.3 baryons per cubic metre (a density compatible with what we observe). The measured abundances could have been all over the place, or could have indicated a mean cosmic density that was plainly ruled out; these nucleo-

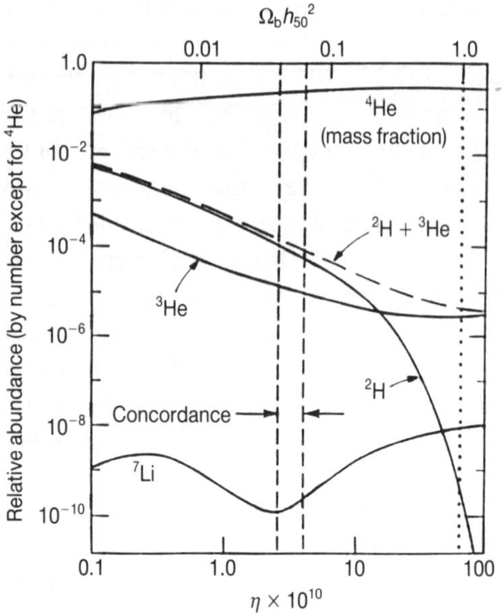

Figure 4 The predicted abundances of the light elements emerging from a standard 'hot big bang', as a function of the baryon/photon ratio η. Note that there is a definite range of η for which the calculations yield abundances of ^{4}He, D, ^{3}He, and ^{7}Li concordant with observations. (From Schramm, D. N. 1991, in *After the First 3 Minutes*, eds. Holt, S. S. *et. al.* (American Institute of Physics, New York) p. 12.)

synthesis calculations therefore offer a strong vindication for extrapolating a standard big-bang model back to a temperature T such that $kT = 1$ MeV. The grounds for this extrapolation should, I believe, be taken as seriously as, for instance, ideas about the early history of our Earth, which are often based on indirect inferences by geologists and paleontologists that are a good deal less quantitative.

Status of the hot-big-bang hypothesis

I would bet odds of 10 to 1 on the validity of the general 'hot-big-bang' concept as a description of how our universe has evolved since it was around 1 second old and at a temperature of 10^{10} K (or ~ 1 MeV). Some people are even more confident. In a memorable lecture at the International Astronomical Union back in 1982, Zel'dovich[18] claimed that the big bang was 'as certain as that the Earth goes round the Sun'. He must even then have known his compatriot Landau's dictum that cosmologists are 'often in error but never in doubt'!

The case for the standard hot big bang has actually strengthened greatly in the last decade, through better measurements of the background radiation and of the light elements. Moreover, one can think of several discoveries that *could* have refuted the model and which have *not* been made. For instance:

(i) Astronomers might have discovered an object whose helium abundance was zero, or at any rate well below 23 per cent. (Stellar nucleosynthesis can readily enhance helium *above* its pre-galactic abundance, but there seems no astrophysically plausible way of eliminating it.)

(ii) The background radiation spectrum might, as experimental precision improved, have turned out to be embarrassingly different from a black body. In particular, the millimetre-wave background measured by COBE might have been *below* a black-body extrapolation of what had already been reliably determined at centimetre wavelengths. It would not be hard to think of effects that would have added extra radiation at millimetre wavelengths (indeed the smallness of the millimetre excess strongly constrains the input from early star formation, decaying particles, etc.), but it would be hard to interpret a millimetre-wave temperature that was *lower* than a black body fitting the Rayleigh–Jeans part of the spectrum.

(iii) If a stable neutrino had been discovered in the mass range

> from 100 to 10^6 eV, that would have been incompatible with the standard big-bang model, which would predict about 1.1×10^8 such neutrinos per cubic metre; relic neutrinos would then provide a far higher density in the present universe than is compatible with observations.

These considerations give us confidence in extrapolating right back to the first few seconds of our universe's history and in assuming that the laws of microphysics were the same then as now. Conceivably, this confidence is misplaced, and our satisfaction will prove as transitory as that of a Ptolemaic astronomer who has fitted a new epicycle. But the 'hot big bang' certainly seems vastly more plausible than any equally specific alternative.

If we envisage time on a logarithmic scale, then many important events of cosmic history are being overlooked if we consider only the period $t > 1$ s. Figure 5 depicts how the universe might have evolved right from the Planck time to the present. Uncertainties about the relevant physics impede our confidence in discussing the extensive span of logarithmic time 10^{-43} s $< t < 10^{-4}$ s, when thermal energies exceeded 100 MeV. At later times, we can consistently use microphysics that is well known; so long as the universe remains almost homogeneous, the evolution is straightforwardly calculable. However, at some stage small initial perturbations must have evolved into gravitationally bound systems (protogalaxies? protoclusters?); the onset of nonlinearity then creates challenging complications, even though the controlling physics is Newtonian gravity and gas dynamics.

Despite the importance and fascination of the ultra-early universe it would be imprudent to venture any bets on what happened when $t \ll 1$ s. The empirical basis for these initial phases of cosmic history is far more tenuous than the quantitative 'fossil evidence' (from light elements and the background radiation) for the eras after 1 second. The first millisecond of

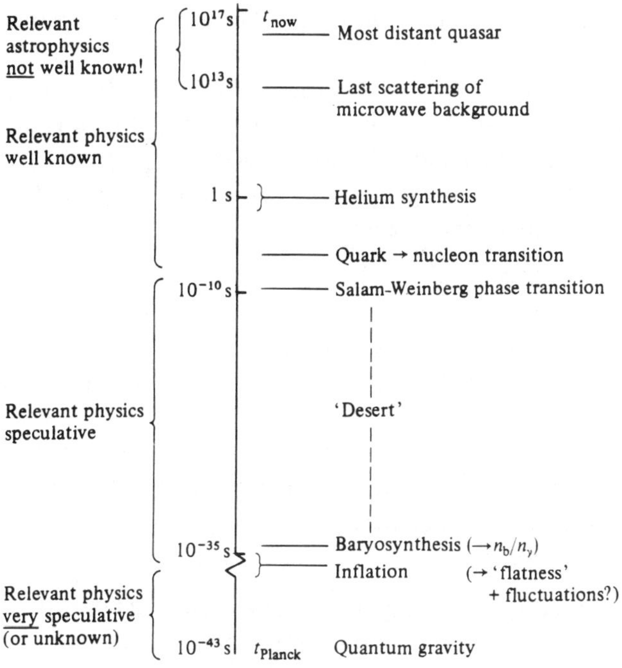

Figure 5 The history of our universe according to the standard hot-big-bang model, showing some of the key physical processes at various stages.

cosmic history, a brief but eventful era spanning 40 decades of logarithmic time (starting at the Planck time), is the intellectual habitat of the high-energy theorist and the inflationary or quantum cosmologist. Densities and energies were then so high that the relevant physics is speculative.

From 10^{-3} seconds onwards, quantitative predictions, such as those about cosmic light-element production, are possible; these vindicate our backward extrapolation. (These predictions also, incidentally, vindicate the assumption that the laws of microphysics were indeed the same when the universe had been expanding for only 1 second as they are in our terrestrial laboratories – something which should indeed be held open to question).

There has been remarkable progress in the last 25 years in delineating cosmic evolution, mapping out the structure and dynamics of clusters and superclusters, and surveying objects at high redshifts. This progress brings new, strongly interrelated, questions into sharper focus:

(i) How did the dominant present-day structures in our universe – galaxies and clusters – emerge from amorphous beginnings in the early universe?
(ii) What is the dark matter that seems to be the dominant constituent of the universe?
(iii) Are the key parameters that have determined the nature of the present-day universe – the structure, the baryon content, the dark matter, etc. – a legacy of exotic physics in the ultra-early phases?

These lectures are mainly concerned with the first two of these questions (which are interlinked), but the final chapter will touch briefly on current conjectures that relate more directly to the still-mysterious physics governing the earliest phases of cosmic history.

2

Galaxies and dark matter

What are galaxies?

Despite the 'broad-brush' homogeneity of the universe, it is the *structures* within it – the individual galaxies themselves, and the groups and clusters into which galaxies are aggregated – which constitute the main subject matter of astronomy. It is a prime goal of our subject, and the main theme of these lectures, to understand how the universe has evolved from an initial dense fireball $\sim 10^{10}$ years ago to its present state, where galaxies dominate the large-scale cosmic scene.

But it is appropriate to start off by focussing on the galaxies themselves. Some of the most basic questions about them are still unresolved. In particular:

(i) We do not know *why* such things as galaxies should exist at all – why these assemblages of stars and gas with fairly standardised properties are the most conspicuous large-scale features of the cosmos.
(ii) About 90 per cent of the mass associated with galaxies is hidden. The luminous stars and gas contribute about a tenth of the gravitating material inferred from dynamical arguments. What the rest consists of is still a mystery.
(iii) It is unclear why the nuclei of some galaxies flare up and release the colossal amount of non-stellar radiation emitted from quasars and radio galaxies.

We are perplexed about these issues, just as 75 years ago our predecessors were perplexed about the nature of stars. But some of us are hopeful that the physical processes underlying galaxies are coming into focus, and can at least be seriously addressed. The following discussion, being short and 'broad brush', is inevitably somewhat distorted, but I hope it will convey the essence of current ideas.

In their classic book on galactic dynamics, Binney and Tremaine[19] assert that galaxies are to astronomy what ecosystems are to biology. They are not only dynamical units, but chemical units as well. The atoms we are made of come from all over our Milky Way galaxy, but few come from other galaxies. The ecological analogy reflects other features of galaxies: their complexity, ongoing evolution, and relative isolation.

Single stars, the individual organisms in the galactic ecosystem, can be traced from their birth in gas clouds through their life-cycle. And we have come to understand why stars exist with the general properties we see. The question why *galaxies* exist is less straightforward than the equivalent question for stars. Galaxies formed at an earlier and remote cosmic epoch. We do not know how much can be explained in terms of ordinary processes accessible to study now, and how much has its causes in the earliest universe.

There is an elaborate taxonomy for galaxies, but the two most obvious categories are discs, and spheroids or ellipticals. A well-known 'cartoon' model, dating back about 30 years, may account for this basic morphological distinction. Suppose that a galaxy started life as an irregularly shaped gas cloud contracting under gravity, and that the collapse were highly dissipative, in the sense that any two globules of gas that collided would radiate their relative kinetic energy and merge (Figure 6). The end result of the collapse would be a rotating disc. This is the lowest-energy state that the gas can reach if it does not lose or redistribute its angular momentum. Stars (or indeed any 'compact' objects) are unlikely to collide with each other, and so cannot dissipate energy in the same fashion as gas clouds. So the *rate of conversion of gas into stars* could be a crucial feature determining the type of galaxy that results.

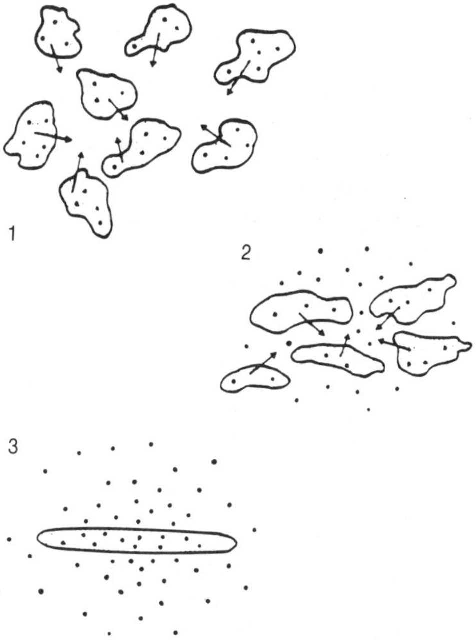

Figure 6 Schematic 'cartoon' showing the collapse of an inhomogeneous slowly spinning gas cloud. The energy of internal gas motions is dissipated during the infall, and the gas eventually settles into a disc. Stars that formed during the collapse would, however, retain their motions perpendicular to the disc and form a 'spheroidal' component.

Elliptical galaxies would be those in which the conversion was fast, so that most stars had already formed before the gas had time to settle into its lowest-energy state. Disc galaxies result when the star formation is delayed until the gas has already settled into a disc. According to this traditional picture, disc galaxies are those with slower 'metabolism': they take longer to approach an asymptotic state in which essentially all the gas is tied up in low-mass stars or dead remnants.

What is special about galactic dimensions?

One serious inadequacy in the foregoing 'explanation' is that Figure 6 has no obvious preferred scale; galaxies do, however, have a characteristic size, even though they have (like stars) a broad luminosity function. Is there any physics that singles out clouds of galactic dimensions, just as, since Eddington and Chandrasekhar, we have understood the natural scale of stars? To some extent, the scales of galaxies must be determined by cosmology – they could not exist unless the 'initial' conditions and dynamics of the expanding universe allowed gravitationally bound gas clouds to condense out. However, there is obviously something which determines where in the mass hierarchy *individual galaxies* end and *clusters of galaxies* begin – why, for instance, is the Coma cluster not a huge amorphous agglomeration of 10^{14} stars?

There is a simple physical argument which suggests at least part of the answer.[20] Consider a set of gas clouds, all held together by gravity but of differing masses and radii. Two timescales are important in determining how a self-gravitating gas cloud evolves. The first of these is the dynamical or freefall time, which is of the order of $(G\rho)^{-1/2}$, its precise value depending on the geometry of the collapse. The second is the radiative-cooling timescale. This depends on the gas temperature T_g and can be written $kT_g/\rho\Lambda(T)$, where $\Lambda(T)$ can be calculated from atomic physics (Figure 7).

If t_{cool} exceeds $t_{dynamical}$, a cloud of mass M and radius r can be in quasi-static equilibrium, the gas being at close to the virial temperature. But if $t_{cool} < t_{dynamical}$, such equilibrium is impossible (Figure 8). The cloud cools below the virial temperature and undergoes freefall collapse or fragmentation. Clouds could collapse and fragment in the fashion depicted in Figure 6 only if they entered the part of the M–r plane where cooling was faster than freefall. A simple calculation shows that this criterion involves a characteristic mass-independent radius of the order

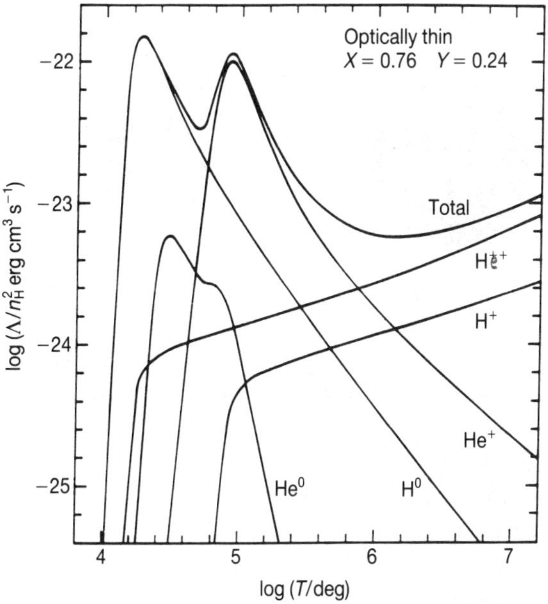

Figure 7 Cooling function for an optically thin plasma with a primordial composition in ionisation equilibrium. This includes thermal bremsstrahlung on hydrogen and helium (H^+ and He^{++}), radiative recombination, dielectronic recombination, and excitation of discrete levels. (From Fall, S. M. & Rees, M. J. 1985, *Astrophys. J.* **298**, 18 (Fig. 1).)

of 75 kpc and a characteristic mass M_{crit} of the order of 10^{12} $M_{\odot}$. Clouds less massive than M_{crit} will readily fragment, but above M_{crit} fragmentation is impossible unless the cloud contracts until its radius is below r_{crit} (Figure 9).

This characteristic mass and radius, consequences of the straightforward atomic physics summarised in Figure 7 combined with the requirements for gravitational equilibrium, feature in many cosmogonic schemes as at least setting an upper limit to the scale of galaxies.

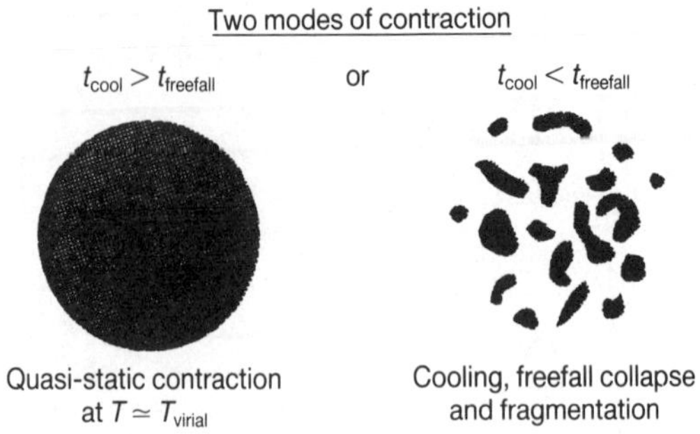

Figure 8 The two collapse modes for a self-gravitating gas cloud, depending on the relative rates of cooling and freefall.

Eddington claimed that a physicist on a cloud-bound planet could have predicted the properties of the gravitationally bound fusion reactors that we call stars. These simple considerations are relevant to galaxies, even though they are not sufficient to 'predict' them. A full explanation of galaxies must involve setting them in a cosmological context. Also, there is the embarrassing fact that most of their mass, maybe as much as 90 per cent, is unaccounted for – it is not in the stars and gas that we see, but takes some unknown 'dark' form.

Dark matter

Galactic halos

One line of evidence for dark matter comes from the discs of galaxies like our Milky Way or Andromeda. These consist primarily of stars, but also contain interstellar clouds of neutral hydrogen (HI). These clouds, themselves just a small fraction of the total mass, serve as a tracer of the orbital motion.[21] Radio astronomers can detect HI, via emission of the hyperfine-structure 21 cm (1420 MHz) line, out at radii far beyond the limit of the optically detectable disc. This outlying gas is in the

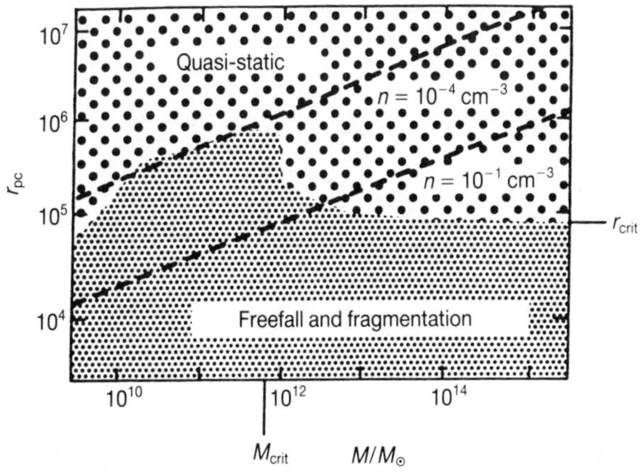

Figure 9 This diagram delineates the mass–radius relation for a self-gravitating gas cloud whose cooling and dynamical timescales are equal (assuming cooling is due only to the processes included in Figure 7). A cloud of given mass whose radius was initially very large would deflate quasi-statically (because $t_{cool} > t_{dyn}$) until it crossed the critical line; it would then collapse in freefall and could fragment into stars. This simple argument (which can readily be modified to allow for non-spherical geometry, a non-baryonic component of mass, etc.) suggests why, irrespective of the cosmological details, no galaxies form with baryonic masses $\gg 10^{12} M_\odot$ and radii $\gg 10^5$ pc. Above $10^{12} M_\odot$, a system has to be compressed to a density $> 10^{-1}$ cm^{-3} before going into freefall collapse; but below this mass, the required density is no more than 10^{-4} cm^{-3}. To fill in the details we need to know more about the initial fluctuations, and also about the efficiency of star formation in protogalaxies.

plane of the disc, and the orbital speed is roughly the same all the way out. If the outermost clouds of HI were feeling just the gravitational pull of what we can see, their speeds should fall off roughly as the square root of distance outside the optical limits of the galaxy: the outer gas would move slower, just as Neptune and Pluto orbit the Sun more slowly than the Earth does. So

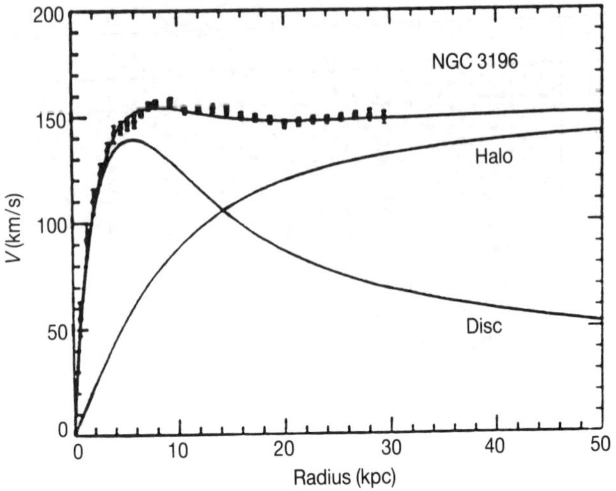

Figure 10 The rotation rate as a function of radius for the disk galaxy NGC 3196. The lower line marked 'disc' indicates the rotation that would be expected if the mass were proportional to the light, which falls off steeply outside 10 kpc. A dark halo must provide the dominant mass at large radii, so that the two contributions (added in quadrature) give the measured 'flat' rotation curve. (From Begemann, K. G. 1989, *Astron. Astrophys.* **223**, 47.)

these rotation curves imply that an extended invisible halo surrounds a galaxy like our own, just as, if Pluto were moving as fast as the Earth, we would have to infer a dark shell of matter outside the Earth's orbit but inside Pluto's[22] (Figure 10).

Further evidence for dark matter within our own Galaxy comes from the orbits of outlying globular clusters and of dwarf satellite galaxies. X-ray astronomers can detect emission from hot gas within galaxies, and determine its density and temperature profile. On the assumption that this gas is gravitationally confined, the depth and extent of the potential well can be deduced; this line of argument corroborates the indications that both disc and elliptical galaxies have extended dark halos, whose density, outside some 'core' several kpc in radius, falls off roughly as r^{-2}. This is the law obeyed by an isothermal sphere, and has the property that the mass enclosed within a radius r goes as the first power of r at large radii.[23]

Groups and clusters

Our own Galaxy and Andromeda both have extensive halos. The total mass of the Local Group (of which these two large disc galaxies are the dominant members) can be derived from the so-called 'timing argument'[24] whereby one calculates how much mass would be needed in order to bring about the present kinematic situation where Andromeda and our Galaxy, now 600 kpc apart, are 'falling' towards each other. These estimates suggest that the two halos may be so massive that they extend out until they effectively merge. The fact that the dynamical timescale in an isothermal halo is proportional to r implies that even in an isolated galaxy the halo must cut off at the radius where this timescale approaches the galaxy's age. Large galaxies without very close neighbours may have typical halo radii of 100 kpc or more. In a cluster, galaxies could not retain individual extensive halos, which would overlap and merge. There then seems, however, to be an equivalent amount of dark matter per galaxy, distributed smoothly through the cluster.

Studies of the dynamics of virialised groups and clusters of galaxies show that these systems would fly apart if they did not contain more mass than is actually seen.[25] The relative motions of the constituent galaxies, and the inferred properties of the hot X-ray-emitting gas pervading the clusters, can be used to determine the mass and density profile of the dark matter, even though they may not themselves have the same density profile.

(The motions of the galaxies in principle trace out the gravitational potential well. But such inferences are difficult in practice because they depend on whether the galactic orbits are isotropic, or else predominantly radial. The gas distribution relative to the dark matter is primarily a function of the ratio of the gas temperature and the virial temperature T. However, problems arise because X-ray astronomers do not yet have high spectral resolution combined with high angular resolution. Moreover the gas may be inhomogeneous and partially supported by bulk motions rather than by thermal pressure.)

An interesting new method of tracing the overall mass distribution in clusters utilises the gravitational deflection of light rays by the cluster. This effect distorts the images of background objects. Some remarkable arc-like objects are observed, which are highly magnified images of faint galaxies lying far behind the cluster.[26] More-detailed studies reveal that *many* background galaxies viewed through rich clusters are somewhat distorted, though less dramatically than the arcs.[27] Such data in principle allow reconstruction of the projected column density of gravitating mass across the entire cluster. This technique can in principle provide a direct estimate of the total amount of (mainly dark) matter, and has already shown that this matter is surprisingly strongly concentrated towards the centre.

The inferred dark matter in galactic halos, groups, and virialised clusters amounts to 10 or 20 per cent of the cosmological critical density $(\frac{8}{3}\pi G t_H{}^2)^{-1}$, where t_H is the Hubble time (i.e. a 'density parameter' $\Omega = 0.1$–0.2). Perhaps the most important issues in extragalactic astronomy are the nature of this dark matter, and whether there is still more of it between clusters – maybe even enough to provide $\Omega = 1$.

What can the dark matter be?

Low-mass stars and VMO remnants

The first and most 'conservative' guess is that the dark matter in our Galaxy is in stars too low in mass to ignite nuclear fuel in their centres. These so-called 'brown dwarfs' or 'Jupiters' would have to be below 7 per cent of a solar mass in order to be faint enough to have escaped detection by conventional astronomy.[28]

Even if they emit no radiation (apart, presumably, from very faint infrared emission) these 'brown dwarfs' can still in principle disclose themselves by gravitationally lensing the light from a background star. The optics of lensing by a single compact object is very simple (Figure 11). The geometry depends on the relative distance of the lens and the background source. However, in the simple case when the lens is roughly half-way along the line of sight to a star at distance d, significant magnification requires alignment to within an angle $\theta = (4GM_l/c^2d)^{1/2}$. The angle θ also characterises the separation of the two images which generally occur in the simple case of lensing by a compact mass. Objects in the halo of our Galaxy could create lensed images of stars in the Magellanic clouds (two small irregular galaxies which are satellites of our own, and 50 kpc from us): θ would then be about 10^{-4} $(M_l/M_\odot)^{1/2}$ arcseconds. For $M_l < M_\odot$, this angle is too tiny for double images to be resolved. One might nevertheless identify instances of lensing by seeing the characteristic rise and fall in magnification when a halo object moves almost across the line of sight to a background star.[29] Unfortunately, even if low-mass stars contributed *all* of our halo's dark matter, the probability of significant lensing, at any instant, along a typical line of sight is only 3×10^{-7}. (This statement is independent of M_l because the lensing cross-section goes as M_l, whereas the number of potential lensing objects needed to make up the halo's dark mass goes as M_l^{-1}.) To stand a chance of detecting an effect, one must therefore keep trying for a very long time indeed or (more optimistically) monitor the lines of sight to millions of different stars. Within the last two years, programmes have started, involving two independent groups, to regularly record the brightness of millions of stars in the Large Magellanic Cloud.

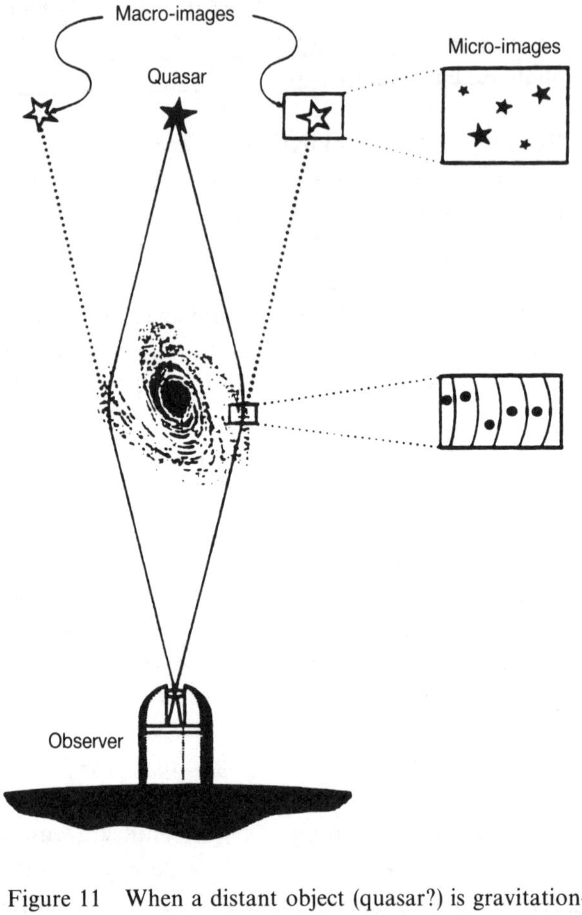

Figure 11 When a distant object (quasar?) is gravitationally lensed by an intervening galaxy, the images are separated by a few arcseconds. If the mass of the galaxy, and its halo, is made up of compact objects of mass M_l, then each image will be 'micro-lensed', displaying structure on an angular scale $\sim 10^{-6}(M_l/M_{\odot})^{1/2}$ arcseconds. (From Refsdal, S. & Sundej, J. 1994, *Rep. Prog. Phys.* **56**, 117.)

Such programmes of course reveal thousands of variable stars of many types: the challenge is to pick out variability which has the characteristic symmetrical rise and fall of a lensing event, and which is also achromatic, in the sense that the amplitude is the same in blue and red light. Both groups have, at the time of writing, just published their first tentative detections of lensing candidates.[30] Within a few more months (certainly before these lectures appear in print), the volume of data analysed should have increased by a factor of 5 or more. There may then be several further (and more convincing) events of the same kind. If so, this will confirm that a new population of low-mass stars is an important constituent of our halo. (On the other hand, if very few further events are discovered, all could be attributed to known populations of stars in the Galactic disc.) The typical transverse speeds and distances of halo objects are calculable, so when there are enough events to constitute a reasonable statistical sample the typical durations should indicate the mass of the lensing objects ($\propto M_1^{1/2}$). Now that the techniques used to distinguish lensing candidates from other kinds of variable stars have been proved to work, the contribution of 'brown dwarfs' to our Galaxy's halo should be settled quite quickly.

Another widely discussed candidate for the dark matter is black holes, which are the remnants of a hypothetical population of very heavy stars that might have formed early in galactic history.[28,31] These so-called VMO (very massive object[31]) remnants would be consistent with most astrophysical evidence if their masses lay in the range 10^3–10^6 $M_\odot$. As already mentioned, the probability that a lensing event is occurring at a given time is independent of the masses of the individual objects making up the halo. However, the detection prospects are less hopeful than for 'brown dwarfs' because the individual events would be much slower and less frequent in the VMO case, and the time one would have to wait before detecting even one scales with $M_1^{1/2}$.

Microlensing at cosmological distances

For a given M_l, the lensing angle θ goes as $d^{-1/2}$, when d is the distance (see Figure 11). So the cross-section for lensing by a given mass, proportional to $(\theta d)^2$, goes *up* as d. Suppose, therefore, that instead of looking at a star in the Magellanic clouds we observe a quasar 5×10^9 parsecs distant – i.e. 10^5 times further away. If a galaxy like our own lay half-way between us and the quasar, the chance that one of the objects in that remote galaxy's halo would lens the quasar is 0.03 (it would be even higher if the line of sight passed through the inner part of the galaxy's halo). The characteristic variability timescale is longer (it scales as $d^{1/2}$), but still only of the order of years for brown dwarfs. So it is in some respects easier to detect lensing by a brown dwarf in a galaxy half-way out to the Hubble radius than to detect one in the halo of our own Galaxy! However, this advantage is outweighed by other difficulties. It is harder to monitor large numbers of quasars than large numbers of stars. Moreover, the intrinsic angular size of even the most compact parts of quasars cannot be assumed to be less than the expected value of θ, which, when d is $>10^9$ pc, is less than 10^{-6} arcseconds for a brown dwarf.[32]

For VMO remnants at cosmological distances, the predicted variability would be too slow to be detectable within our lifetimes. However, lensing by such objects (for which θ is in the range 10^{-3}–10^{-4} arcseconds) could reveal itself by distortion of background radio sources, which can be mapped by VLBI (very long baseline interferometry) techniques with better than milli-arcsecond resolution.[33,34]

Non-baryonic matter

VMO remnants, like 'brown dwarfs', count as *baryonic* matter. Standard primordial nucleosynthesis (cf. Figure 4) pins down the value of the baryon-to-photon ratio η, and thereby sets an upper limit of $0.1h_{50}^{-2}$ to the contribution that baryonic matter makes to Ω, where h_{50} is the Hubble constant in units of 50 km s^{-1} Mpc^{-1}. It would be marginally consistent with this constraint for the halos of individual galaxies to be baryonic; on the other hand, the amount of dark matter inferred in 'superclusters' (though less well pinned down observationally – see Chapter 3) probably exceeds this limit even for the lowest credible values of h_{50}. If Ω were anywhere near unity, the 'standard' model for primordial nucleogenesis would need drastic revision if most of the dark matter were baryonic. (Certain 'non-standard' models, where the baryons are postulated to be clumped on scales smaller than the neutron diffusion length at the time the nuclear reactions occur, may allow a slightly higher Ω_b. However, the early hopes[34] that these more complicated models would allow a much wider 'window' of acceptable densities have not been borne out;[35] nor is there much physical motivation for the assumptions underlying them.)

Let us therefore turn to the *non*-baryonic candidates – physicists are likely to find these more interesting anyway. The most obvious option is neutrinos. If the big bang started off with $T > 10^{11}$ K, neutrinos would have come into thermal equilibrium with photons. The present density of each neutrino species would then be 3/11 the photon density. This fraction is the product of two factors: 3/4 because neutrinos are fermions rather than baryons, and 4/11, which is the reciprocal of the factor by which electron–positron annihilation (occurring when T falls below $m_e c^2/k = 5 \times 10^9$ K) would have enhanced the photon density. Each species of neutrino should therefore have a mean number density of 1.1×10^8 m^{-3}. (Actually, this is only true if the masses are below about 1 MeV; otherwise the mass term in the Boltzmann factor would have reduced the number before they decoupled from the photons and electron–positron pairs.) Because neutrinos outnumber baryons by such a big factor in a 'hot' big bang, they would not need a very high rest mass in order to have an important cumulative effect on cosmic dynamics: they would contribute the full 'critical' density for a mass of $23h_{50}{}^2$ eV.

Twenty years ago, Cowsik and McClelland[36] and Marx and Szalay[37] conjectured that neutrinos could provide the dark mass in galactic halos and clusters. At that time the suggestion was not followed up very extensively. But by the 1980s physicists had become more open-minded about non-zero neutrino masses. (Moreover, there was an experimental claim by Lyubimov *et al.*,[38] generally now discounted, that the electron-neutrino mass was 36 eV.) Theorists were then stimulated to explore scenarios for galaxy formation in which neutrino clustering and diffusion played key roles (see Chapter 3). Firmer clues to neutrino masses may soon come from terrestrial-beam experiments, or from studying solar neutrinos.

Neutrinos have the virtue of being known to exist. But theoretical physicists have a long shopping list of hypothetical particles that might have survived in the requisite numbers from the early universe. One widely discussed possibility is weakly interacting massive particles (WIMPs) predicted by supersymmetric theories. If WIMPs were the dominant dark matter in our own Galactic halo, their local density would be around 0.1 $(m_w/m_{proton})^{-1}$ cm^{-3}, and they would be moving with typical halo speeds of $\sim$300 km s^{-1}. Attempts are underway, by several groups, to search for WIMPs by detecting the recoil in the rare events when one of these particles interacts with a nucleus. The (weak-interaction) cross-sections are very low; the predicted rates, dependent on specific assumptions and on the target material, are typically only a few per day per kilogram of detector. These searches are bedevilled by extraneous background from internal radioactivity of the equipment, and from muons created by cosmic rays; experiments must be carried out deep underground. There is fortunately one unambiguous diagnostic which could distinguish a genuine WIMP-induced signal from other background events. The predicted interaction rate due to WIMPs is sensitive to velocity; the mean event rate would have an annual variation, because as the Earth moves round the Sun its velocity relative to the halo changes. The expected annual modulation, with a peak in June and a minimum in December, and an amplitude of a few per cent, would unambiguously distinguish WIMP-induced events from other backgrounds. Even an optimist would not rate the chance of positive results from such experiments as being as much as 50 per cent. But the goal is worth shooting for because success could not only tell us what most of the universe was made of, but also reveal a new class of elementary particle.[39]

Axions are another widely discussed candidate, and there is even experimental interest in searches for these, via conversion into photons or interaction with matter or strong magnetic fields. The prospects seem rather more discouraging than for WIMPs, because the photons would be in a narrow energy band, depending on the poorly known axion mass, so a broad range of photon energies (spanning the millimetre and infrared bands) would need to be searched.[39]

How to discriminate among dark-matter options

It is not surprising that most of the matter in the universe should be dark: there is no reason why everything should shine, and it is not difficult to think of possible candidates. The problem is still to discriminate among a rather long list of possibilities. Obviously *direct detection* is the 'cleanest' and most decisive discriminant: dark stellar objects in the Galactic halo can reveal themselves by gravitational microlensing; exotic particles which pervade the galaxy (and therefore are continually passing through every laboratory) may be detected by sensitive experiments.

The late Professor Redman of Cambridge, a 'no-nonsense' observer with little taste for speculation, once claimed that a competent astrophysicist can reconcile any theory with any new observation. An even more cynical colleague extended this claim, asserting that the astrophysicist often need not even be competent. Dark-matter theorists have in the past sometimes exemplified Redman's 'theorem', and its extension as well. But searches may soon yield definitive positive results; in any case, various constraints are now restricting the tenable options. It is not, however, 'wishful thinking' to expect that there may be more than one important kind of dark matter – for instance, non-baryonic dark matter could control the dynamics of large clusters and superclusters, even if individual galactic halos contained a lot of 'brown dwarfs' or VMOs.

It would be specially interesting if astronomical techniques were to reveal some fundamental particle which has been predicted by theorists. If such particles turned out to account for the dark matter, we would have to view the galaxies, the stars, and ourselves in a downgraded perspective. Copernicus dethroned the Earth from any central position. Early this century, Shapley and Hubble demoted us from any privileged location in space. But now even *baryon chauvinism* might have to be abandoned: the protons, neutrons, and electrons of which we and the entire astronomical world are made could be a kind of after-thought in a cosmos where photinos or neutrinos control the overall dynamics. Great galaxies could be just a puddle of sediment in a cloud of invisible matter ten times more massive and extensive.

Apart from direct detection, two other lines of attack could narrow down the possibilities for dark matter.

(i) Progress in basic physics will obviously help. We do not yet really know what types of (supersymmetric?) particle might be present in the ultra-early phases of the universe, nor what their cross-sections are for annihilation. But if we knew the mass and annihilation cross-sections for any species (together with any possible 'favouritism' of particles over antiparticles), we could in principle calculate how many would survive. When/if theoretical advances or accelerator data eventually tell us this, the number of such particles surviving into the present universe (and their contribution to Ω) will become as confidently predictable as the proportions of deuterium and helium are today.

(ii) We can compute the distinctive implications of each hypothesis for cosmogony. Since the dark matter is gravitationally dominant, the large-scale structure in the universe, and perhaps the morphologies of the galaxies themselves, are essentially determined by how it aggregated under gravity as the universe expanded. In this context, the non-baryonic options divide into two main categories – 'hot' and 'cold'. Neutrinos of 10–30 eV come in the 'hot' category, in the sense that, though they are now moving slowly, their thermal motions were sufficiently high during earlier phases for free-streaming to have homogenised fluctuations on small scales. 'Cold' candidates, on the other hand, of which supersymmetric particles and axions are the prime examples, would never have had significant thermal velocities (except at times $\ll 1$ second), and exert no significant pressure; they can therefore agglomerate into bound systems on all scales where primordial fluctuations exist as 'seeds' for gravitational instability. The emergence of cosmic structure is the subject of the next chapter.

3

Emergence of cosmic structure

Gravitational instability

When pressure gradients are unimportant, linear fluctuations increase their density contrast as the universe expands. If $\Omega = 1$, the growth factor in the linear regime is exactly proportional to the scale factor R; for other values of Ω, the growth of linear perturbations 'saturates' at recent epochs, though the amplitude is still proportional to R when $R/R_{\rm now} < \Omega$. (These results follow from simple Newtonian arguments applied to an overdense sphere or sine-wave perturbation: the growth follows a power law, rather than being exponential, because the background universe is expanding on the same timescale, $(G\rho)^{-1/2}$, as that on which the perturbation grows.) Any regions whose excess density at recombination substantially exceeded 10^{-3} would, by the present time, have become nonlinear and evolved into bound systems. We can understand why $(\delta\rho/\rho) \propto R$ in another way by noting that, when pressure gradients do not act, the metric fluctuation (or energy deficit per unit mass) associated with the perturbation, proportional to $GM^{2/3}$ $(\delta\rho/\rho)$ R^{-1}, must remain constant during the expansion.

Figure 12 shows how the radii of overdense spheres behave during the expansion of the universe, assuming pressure can be neglected. If the initial overdensity is large, expansion is halted (and the fluctuation 'goes nonlinear') at an early stage, and there is ample time for the system to recollapse and establish a virial equilibrium. If the initial amplitude is rather smaller, the sphere may by now have stopped expanding, and commenced its infall, without yet having virialised. And a sphere with still

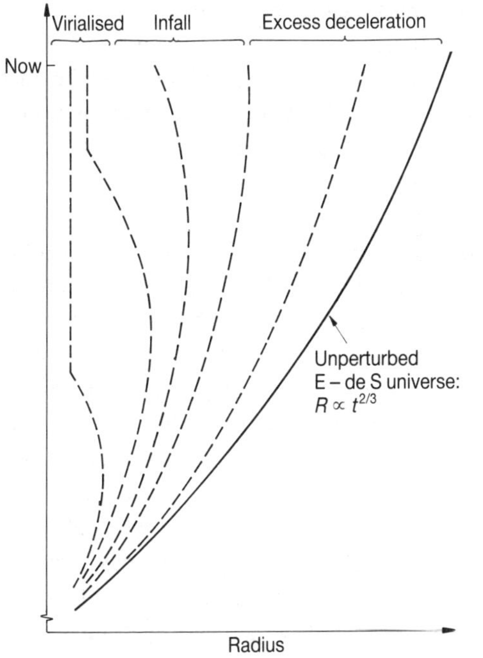

Figure 12 The dynamics of overdense spheres with $p = 0$ in the expanding universe. The larger the initial overdensity, the earlier the sphere's expansion halts. Systems which are already virialised must have $\rho/\rho_{\text{crit}} \gtrsim 200$; those which are now infalling must have $\rho/\rho_{\text{crit}} \gtrsim 5$. Note that the overdensity factor of bound or collapsing systems must be even larger in a universe with $\Omega < 1$ and mean density below ρ_{crit}. Realistic perturbations would of course not be spherical.

smaller initial overdensity would still be expanding, though it would have suffered an excess deceleration, and its constituent particles would therefore not be moving exactly with the mean Hubble flow. In the simple case of spherical perturbations in an Einstein–de Sitter universe, any already-virialised system must have a present density more than 200 times the mean. A system which has halted its expansion and is now displaying infall must have more than 5 times the mean cosmic density. (A sphere is of course an unrealistic special case. For the rather more general case of an ellipsoid, collapse along one axis can begin while expansion still continues along the others: there is no time at which the kinetic energy is zero, and therefore such systems do not have to contract by as much as a factor of 2 to establish virial equilibrium.)

The simple dynamics depicted in this figure is relevant in two different, but related, contexts. If we imagine a single sphere, which condenses around a central high-density peak, the dashed lines can represent different shells: the inner ones feel a large fractional overdensity, and collapse early; the outer ones feel only a small perturbation, and therefore are merely slightly decelerated. If the early universe contained a spectrum of initial fluctuations, such that $(\delta\rho/\rho)$ fell off towards larger scales, then we could also use the same figure to infer that smaller systems would tend to have already virialised, whereas larger scales, which initially had smaller density enhancements, would be 'dynamically younger', and would not yet have achieved dynamical equilibrium. More realistic cosmogonic models, which postulate random fluctuations with a specific spectrum, can be simulated by N-body calculations.

It is perhaps helpful to define a *cluster* as a gravitationally bound system which, at least in its centre, has achieved virial equilibrium; and a *supercluster* as a larger system which, even though it may contain virialised substructure, is overall in a dynamically younger state – perhaps even still expanding with the universe, albeit at a decelerated rate.

Most galaxies are not in large clusters. A simple measure of clustering is the 'two-point correlation function', which measures the excess probability of finding a second galaxy at a given distance from a first.[40] This function drops below unity at distances above about $16h_{50}{}^{-1}$ Mpc, and there are only a few galaxies in a typical region with that diameter. The rich clusters, in any gravitational-instability scenario, would have evolved from regions where the initial fluctuation amplitude on the relevant mass-scale was exceptionally large (e.g. a $>3\sigma$ peak for fluctuations whose amplitude distribution is Gaussian).

The conspicuous cosmic structures – galaxies, clusters, and superclusters – would, in the early compressed phases of cosmic expansion, have been merely slightly overdense regions, which lagged behind and eventually separated from the cosmic expansion. There must have been some primordial fluctuations for the clusters to grow from – otherwise the universe would still now be cold hydrogen, with no galaxies, no stars, and no people. Over the last few years there have been elaborate simulations of how this gravitational clustering proceeds. The linear fluctuations that are 'fed in' at the start of the simulation depend on the cosmological assumptions – in particular, on whether the dark matter is 'hot' or 'cold', and also on whether the fluctuations are Gaussian.

The fluctuation spectrum at t_{rec}

At (re)combination, when the universe was $\sim 3 \times 10^5$ years old and had cooled to ~ 3000 K, the primordial black-body radiation shifted redward of the visible band, and the universe literally entered a dark age which persisted until the first bound systems condensed and started to radiate. We do not know when 'first light' occurred. The dark age may have been rather brief; on the other hand, it could have lasted for about 10^9 years, almost until the epoch of the high-redshift quasars. We remain more confused and ignorant about this phase of cosmic history than many seem to be about the first 10^{-35} seconds.

The key determinant of the cosmogonic process is the spectrum of density fluctuations – the r.m.s. amplitude as a function of mass-scale – at t_{rec}. Perturbations on all scales above $10^6 M_\odot$ amplify at the same rate thereafter (growth of smaller-mass perturbations is impeded, even after the radiation has decoupled, by the small pressure of the baryons themselves).

The first bound systems to arise via gravitational instability will have mass-scales for which the fluctuation amplitude peaks. If non-baryonic matter makes the dominant contribution to Ω, then it plays the dominant role in the gravitational aggregation of protogalaxies and clusters.

The spectrum depends on what is imprinted initially, possibly modified by preferential damping of smaller scales before recombination. Three examples are shown in Figure 13, the amplitude at t_{rec} being normalised on a mass-scale of $10^{15}\ M_{\odot}$ so as to give the appropriate present-day amplitude on this scale (see caption for further explanation).

The right-hand panel of Figure 13 shows a 'white-noise' spectrum, with amplitude increasing as a straight power law towards smaller scales. Here we have a hierarchical 'bottom–up' cosmogony, with the emergence first of sub-galactic scales (of $10^6 M_{\odot}$), then galaxies, and then clusters. (There may then be an interesting complication: radiative or explosive output from the first small bound objects could create secondary large-scale inhomogeneities that swamp those already present). Because scale-dependent physical processes in the earlier universe ($t < t_{\text{rec}}$) would modulate any spectrum, there is no natural physical model which would lead to a straight power law of this kind at t_{rec}. The other two panels in the figure show the spectra which are expected if the dominant matter is non-baryonic.

The left-hand panel in Figure 13 shows the spectrum expected at t_{rec} if the universe is dominated by neutrinos with masses $m_\nu = 10$–20 eV. When the cosmic temperature was higher than $m_\nu c^2/k$ (i.e. when $T > 1$–2×10^5 K in this example), the neutrinos would be moving relativistically; but at later stages the thermal speeds of the neutrinos would slow down, decreasing as R^{-1} (i.e. in step with the photon temperature). Because of the high thermal speeds before t_{rec}, the neutrinos would homogenise[41] on scales up to at least $10^{14}\ M_{\odot}$. The first bound systems would then be superclusters, and galaxies would result from some kind of secondary fragmentation process.

Let us focus now on the middle panel in Figure 13. The fluctuation spectrum here has the shape expected if the dominant matter is WIMPs, or any non-baryonic material that is 'cold', in the sense that the individual particles move too slowly for damping due to free-streaming to occur, as it does for neutrinos. The spectrum is calculated[42] on the further assumption of a 'Harrison–Zel'dovich'[43] initial spectrum. This spectrum is 'natural' in that the postulated metric fluctuations in the early universe had the same amplitude ψ on all scales. This means that, for a perturbation on the scale of the 'Hubble radius' ct, $(\delta\rho/\rho)$ would be equal to ψ. However, a 'Hubble volume' encompasses less mass at earlier times: indeed, since $(G\rho t^2)$ is constant in the early stages of a Friedmann universe, $\rho(ct)^3$ is clearly proportional to t. The Harrison–Zel'dovich spectrum implies that perturbations all have the same amplitude ψ when they occupy a 'Hubble volume'. This happens earlier for low masses. If growth were thereafter unimpeded by pressure, so that $(\delta\rho/\rho) \propto R$, then the smaller mass-scales would have had a 'head start' in their growth, and at any epoch the density fluctuation amplitude would fall off towards larger scales as $M^{-2/3}$.

In the case of cold dark matter, the spectrum at t_{rec} is not a simple power law: at very large scales $(\delta\rho/\rho) \propto M^{-2/3}$; but towards smaller scales it gradually 'rolls over' towards a spectrum where $(\delta\rho/\rho)$ is almost independent of M. The reasons for this are summarised in Figure 14 and its caption. The rollover occurs on a mass-scale equal to that within a 'Hubble volume' at the time when radiation and matter have equal density (corresponding to a cosmic epoch when $(R/R_0)^{-1} = 10^4\,\Omega\,h_{50}{}^2$). In fact, as is clear from Figure 15, a full calculation shows that the bend in the spectrum is very gradual. The 'vertical' normalisation is not determined by the theory. When the normalisation on a mass-scale $10^{15} M_\odot$ is chosen to fit the present-day clusters of galaxies, the spectrum implies that a

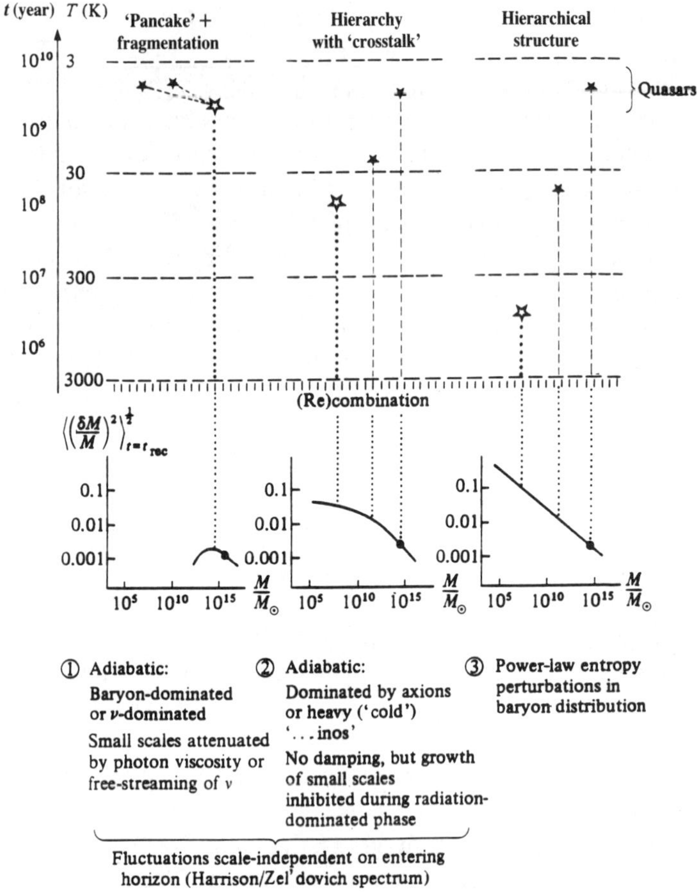

Figure 13 The cosmogonic processes after (re)combination depend on the spectrum of fluctuations which have survived damping processes, etc., at earlier times. After t_{rec}, linear growth proceeds roughly according to the law $(\delta\rho/\rho) \propto t^{2/3}$, and the first gravitationally bound systems to form will have the mass for which the density contrasts at t_{rec} are biggest. The epochs of collapse are indicated by 'stars' in the upper part of the diagram. In case (1), which resembles what is expected if the baryon-to-photon ratio is uniform but the

typical fluctuation of $10^6 M_\odot$ would collapse no earlier than the epoch corresponding to $z = 10$. The build-up of structure is hierarchical, in the sense that smaller masses tend to virialise earlier, and thereafter become subsumed in larger systems. However, because of the flat spectrum, there would be complicated 'cross-talk' between many different scales. The 3σ peaks in the density distribution on galactic scales, $10^{11} M_\odot$, would have similar amplitudes to more typical peaks of mass $10^6 M_\odot$, and would therefore collapse at about the same time. It is consequently hard to analyse, either analytically or numerically, even the purely dynamical and non-dissipative aspects of the clustering. However, those studies that have been done are encouraging,[44,45] in that when the amplitude of the fluctuations is normalised so as to match the data on galaxy clustering, the finer-scale disposition of the dark matter closely reproduces the sizes and profiles of individual galactic halos. An example of

---

Caption for Figure 13 (*cont.*)

dark matter is in light neutrinos, (super)clusters are the first systems to condense, and they form quite recently. Models of this kind run into difficulties because we observe quasars with redshifts z as large as 5, whereas if superclusters had collapsed as early as this, they would now be denser, and display a higher density contrast, than is seen. In case (2), baryonic systems of $\sim 10^6 M_\odot$ condense in potential wells produced by 'cold-dark-matter' particles, which are presumed to be slow moving, so that they are not homogenised on small scales as neutrinos are. (This specific case is discussed more fully in the text.) The third case shows the spectrum that might arise from an *ad hoc* power-law spectrum of perturbations in the baryon-to-photon ratio. In cases (2) and (3), sub-galactic systems would form *before* galaxies; if these sub-galactic systems provided an energy input, they could in principle generate 'secondary' perturbations on larger scales which could swamp the genuinely primordial ones.

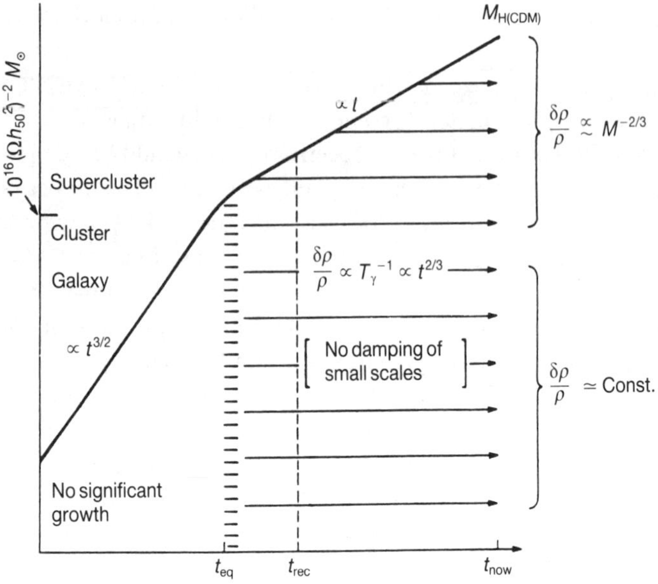

Figure 14 The growth of adiabatic fluctuations in a universe dominated by 'cold dark matter'. The mass of cold dark matter within the horizon scale *ct* is shown as a function of time on a log–log plot. For $t > t_{eq}$ (corresponding to the redshift indicated on the figure), all scales grow at the same rate. Before t_{eq}, when the expansion is radiation dominated, there is essentially no growth, because the growth timescale much exceeds the expansion timescale. If the fluctuations on all scales enter the horizon with equal amplitude ψ (the 'Harrison–Zel'dovich hypothesis'), then the present-day spectrum would have the approximate form written on the right-hand side. (The accurately calculated spectrum is shown in Figure 15.) The CDM perturbations start to grow at t_{eq} whereas radiation pressure inhibits growth of baryonic fluctuations on relevant scales until the (later) recombination time t_{rec}. CDM fluctuations thus have a 'head start' (baryons being able to fall into the resultant potential wells after t_{rec}); this permits an acceptable cosmogonic scheme with lower fluctuation amplitude ψ, and smaller microwave-background fluctuations, than in a baryon-dominated universe.

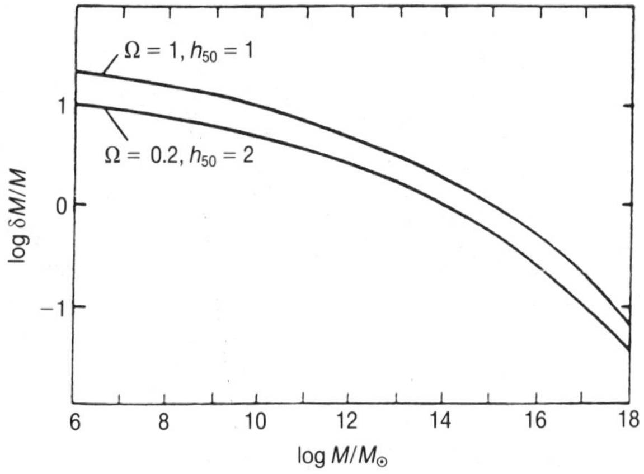

Figure 15 The r.m.s. fluctuations, as a function of M, for two CDM models. (Adapted from Blumenthal *et. al.* 1984, *Nature* **311**, 517.)

how such clustering develops, based on simulations of Frenk *et al.*,[46] is shown in Figure 16.

Is the universe flat?

In an influential review published back in 1974, Gott, Gunn, Schramm, and Tinsley[47] summarised the arguments bearing on Ω. They concluded that the dynamical evidence (from clusters of galaxies etc.) favoured a value of 0.1 or 0.2, and noted that if this matter were all baryonic, the lower end of the range was compatible with the value favoured by big-bang nucleosynthesis, for $h_{50} = 1$ and $t_H \approx 2 \times 10^{10}$ years (a value consistent with the estimated ages of the oldest stars etc.). Much new evidence has accumulated since 1974, especially on cluster dynamics and element abundances, and some relevant theoretical issues have been refined and elaborated. But if one were to update Gott *et al.*'s discussion, their net conclusion would not change much. The attitudes of theorists, however, seem to have

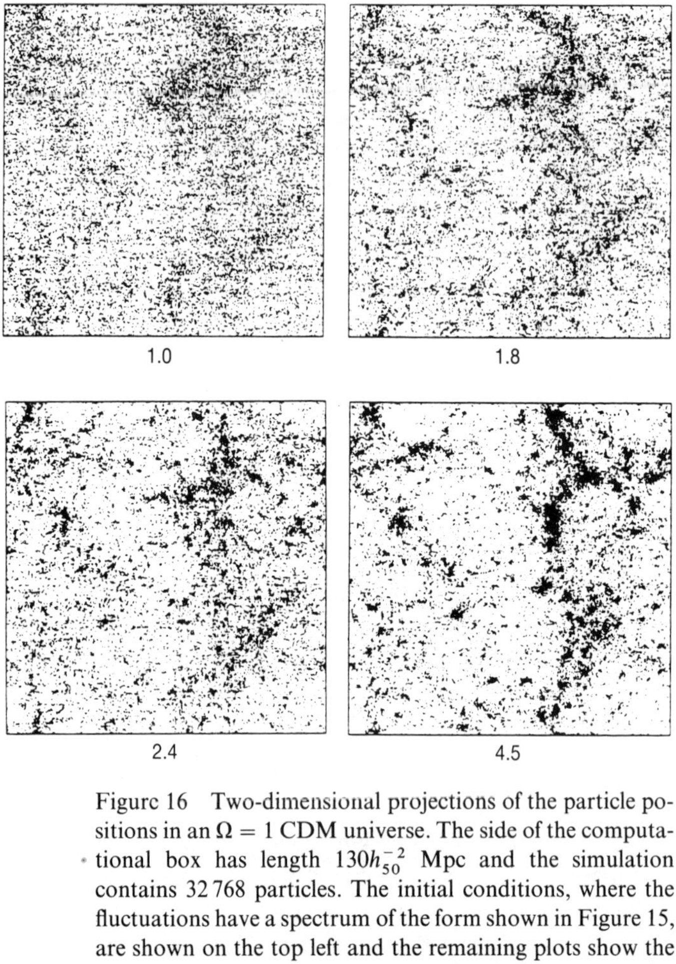

Figurc 16 Two-dimensional projections of the particle positions in an $\Omega = 1$ CDM universe. The side of the computational box has length $130h_{50}^{-2}$ Mpc and the simulation contains 32 768 particles. The initial conditions, where the fluctuations have a spectrum of the form shown in Figure 15, are shown on the top left and the remaining plots show the model after expansion by factors of 1.8, 2.4, and 4.5. (From Davis, M. *et. al.* 1985, *Astrophys. J.* **293**, 371.)

changed markedly. This is partly because non-baryonic matter is now taken much more seriously, seeming in some ways almost a natural expectation. But the other new element in the discussion is the concept of inflation. This seems to resolve some stubborn paradoxes in a rather natural way (see p. 114). In particular, it suggests why the expansion rate is so fine-tuned that our universe has neither collapsed long ago, nor is expanding too fast for galaxies to have condensed. It suggests, moreover, that Ω has almost exactly the value unity. If Ω were indeed 1, the balance of argument would shift strongly in favour of non-baryonic dark matter, 'hot' or 'cold', because big-bang nucleosynthesis favours a value of $\Omega_b < 0.1 h_{50}{}^{-2}$ (see Chapter 1).

As we have seen in Chapter 2, the most clear-cut dynamical evidence for dark matter – from galactic halos and virialised clusters – provides no evidence that Ω exceeds 0.2. Could there, however, be still more matter thinly spread between the clusters? To answer this question required mapping, and quantifying the dynamics, of the huge superclusters, 'great walls', and other large-scale structures.

The last few years have seen rapid progress in delineating these structures. We now realise that the distribution of galaxies in space is significantly structured even on scales as large as $50h_{50}$ Mpc. The well-known Lick counts,[48] whose clustering properties were analysed extensively by Peebles and his co-workers, are now complemented by data from the UK Southern Sky Survey. Maddox and Efstathiou,[49] using the APM (Automatic Plate-Measuring Machine) in Cambridge, have studied the galaxy correlation function and clustering data for this survey. More important still have been the programmes to obtain *redshifts* for many thousands of galaxies, and thereby to map out the structures in three dimensions.[50]

Astronomers react to the large-scale clustering rather as to ink-blot psychological tests. Some see filamentary features, bubbles, or sheets; others see only the outcome of Gaussian fluctuations. Statistical methods applied to the data include percolation statistics, 3- and 4-body correlation functions, the topological characteristics of equidensity surfaces, etc. Better methods are needed not only to characterise the growing body of data, but also to describe the outcomes of numerical simulations. It is, for instance, clear to the eye that the 'hot'-dark-matter simulations have a smaller 'dynamic range' of interesting-looking structure than the CDM (cold dark matter) models, and seem a worse representation of the actual data, but one would like to be able to make these assessments more quantitatively.

The 'supercluster' systems are not virialised, and are generally expanding; the gravitational effect of their excess density should nevertheless generate velocity perturbations which are in principle measurable. For spherical perturbations (cf. Figure 12) where the overdensity is not yet large enough for turnaround to have occurred, the 'peculiar velocity' (that is, the deviation from the Hubble-flow velocity at the same place) is related[51] to the overdensity, and to Ω, by

$$\delta V/V_{\text{Hubble}} = f(\Omega)(\delta\rho/\rho), \tag{1}$$

where $f(\Omega) \approx \Omega^{0.6}$. The overdensity factor $(\delta\rho/\rho)$ may, however, not be equal to the enhancement in luminous galaxies – indeed it is customary to regard these quantities as related by a 'bias factor' b.

Our local 'peculiar velocity' relative to the microwave background is generally thought to have been induced by inhomogeneities in the distribution of galaxies around us. Our Local Group is pulled towards clusters, and (in effect) 'pushed' away from voids. Significant contributions to our peculiar velocity certainly seem to come from distances up to $100h_{50}{}^{-1}$ Mpc away (the 'Great Attractor'[52] region); it is still a matter of debate whether the distribution of matter around us remains sufficiently asymmetric to contribute to our motion even beyond that distance.[53,54]

Since gravity and light both obey the inverse-square law, we can readily infer that the contribution made by any particular galaxy or group is proportional to the light we receive from it, provided that b is indeed a constant. Note also that the gravitational-potential fluctuation ($G\delta M/r$) required to cause a given peculiar velocity ($\propto(\delta M/r^2)$) is larger if the scale is larger – this consideration constrains the amount of our Local Group's peculiar velocity that could be due to structures even further away than the Great Attractor.

Equation (1) can be applied to our own local motion just from two-dimensional data, without knowledge of the redshifts of galaxies. The next step, however, is a much more ambitious one, namely to apply it to the peculiar motions of many other galaxies as well, so as to infer the mass distribution throughout our local region. The peculiar velocities of other galaxies of known redshift can of course only be determined if one has an independent distance indicator that is sufficiently precise to distinguish between a galaxy's *actual* distance and the distance it *would have had* if it were following the undisturbed Hubble flow. The main practical uncertainties in the measurement and interpretation of large-scale streaming stem from the problems of calibrating the distance indicators.

For many galaxies, distances can be estimated accurately enough to determine a peculiar velocity (or at least the component of this velocity along the line of sight). Speeds of several hundred kilometres per second are found. If these large-scale streaming motions ('cosmic plate tectonics') are induced by gravity, one can reconstruct the density field from them, treating the galaxies themselves as 'test particles'.[55]

A year or two ago, many of us were still dubious about these streaming velocities, because of lingering uncertainties about whether the luminosity calibrators (velocity dispersion, diameter, etc.) are really independent of location and environment. But now the evidence hangs together much better, at least in some parts of the sky. The motions of the galaxies, relative to the Hubble flow, indeed seem to be converging towards regions where there is a high concentration of galaxies (see Figure 17 and caption). If the inferred distances were incorrect, and the claimed peculiar velocities spurious, there would be no reason

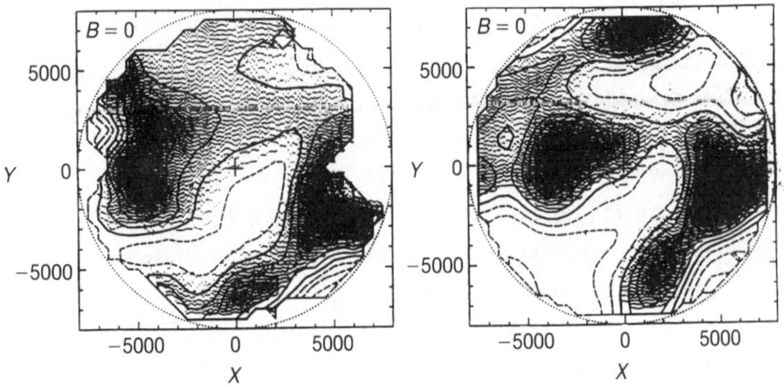

Figure 17 The right-hand panel shows the density contours for the distribution of IRAS (Infrared Astronomy Satellite) galaxies in a region extending out to Hubble recession velocities of 5000 km s^{-1} (a region about half the extent of that shown in Figure 1). The contours in the left-hand panel show the density of matter that would be required in order to induce the observationally determined deviations from the Hubble flow. The obvious resemblance between the two sets of contours indicates that the galaxies are tracers of the overall mass distribution; it also bolsters our confidence that the large-scale streaming velocities indeed exist, rather than being artifacts arising from large-scale environmental effects on the distance indicators. These motions offer the first real dynamical evidence for a mean cosmic density as high as the 'critical' value ($\Omega = 1$). (From Dekel, A. *et. al.* 1993, *Astrophys. J.* **412**, 1.)

to expect them to delineate a dynamically plausible flow. The large-scale motions and their implications are still a subject of controversy, but the evidence is starting to vindicate the distance calibrators which are generally used.

The streaming motions are important because they directly probe inhomogeneities in the gravitational potential, which can be due to the dark matter. The galaxies are test particles and we do not need to assume that they are distributed like the dark matter itself.[56] Gravitationally induced deviations from the Hubble flow that are substantial and widespread (i.e. not restricted to regions around virialised clusters) are incompatible with a low Ω.

These large-scale motions definitely indicate a value of Ω higher than 0.2, and even seem (gratifyingly to inflationary cosmologists) to be consistent with $\Omega = 1$. There is therefore even more dark matter, spread out on the scale of superclusters. We still do not know what it is. However, the more there is of it, the less likely it is to be made of ordinary baryons, since the latter is constrained by big-bang nucleosynthesis calculations to give Ω_b no more than $0.1 h_{50}{}^{-2}$. The most likely non-baryonic options are light neutrinos or cold-dark-matter particles.

What we really need is some technique for delineating the distribution of dark matter on supercluster scales without any assumption about whether galaxies trace mass, or about the universality of the distance indicators needed to determine galactic peculiar motions. One possibility which is now being implemented involves searches for gravitational-lensing distortion of the images of faint background galaxies by superclusters.[57] This is analogous to the method already being applied to clusters (see p. 26); however, a supercluster would not have such a high column density as the core of a rich cluster, and therefore could not give strongly distorted or magnified images. But the angular diameter of a supercluster with redshift (say) 0.3 may be 1 degree; up to 10^5 faint galaxies lying beyond it would share a correlated distortion, so it is by no means hopeless to seek distortions correlated over the whole area viewed through the same supercluster, even if the effect amounts to only a few per cent.

Classical methods for determining Ω

Deceleration parameter

The most direct way of estimating Ω (and of detecting the effect of dark matter even if it is widely dispersed through intergalactic space) is to detect its effects on the curvature and dynamics of the entire universe. Measurement of the 'deceleration parameter' $q = -R\ddot{R}/\dot{R}^2$ (equal to $\Omega/2$ in the simple Friedmann models with cosmological constant $\Lambda = 0$) by extending Hubble's magnitude–redshift relation to high redshifts has been an astronomical goal ever since the 1950s. The prospects, and the basic theory, were laid out clearly in, for instance, Sandage's classic 1961 paper[4] on 'The ability of the 200 inch telescope to discriminate between selected world models'. But from the 1970s, especially through the influential work of Tinsley, it became clearer that evolutionary corrections were significant; moreover galactic evolution would need to be well understood before they could be quantified. The galaxies seen at large distances are systematically younger than nearby ones. Even if galaxies of a certain type (e.g. the brightest galaxies in clusters) constitute precise 'standard candles' at the present, one needs to know how each candle changes as it burns. The galaxies that are crucial for cosmological tests are those whose light has been journeying towards us for 5–10 billion years, which are being seen at less than half their present age.

In a younger elliptical galaxy many stars would be shining which by now have died; and the present stars would all be at earlier stages in their evolution. This alters the brightness and colour, the trend generally being that younger galaxies would be brighter and bluer. The effect depends quantitatively on the mass distribution of the stars, and the history of star formation. But there is a second evolutionary effect stemming from the fact that a galaxy is not a self-contained isolated system. We can see many instances where galaxies seem to be colliding or merging with each other; in rich clusters the large central galaxies may be cannibalising their smaller neighbours. (In a few billion years this may, incidentally, happen in our own Local Group. The Andromeda galaxy is falling towards our own Milky Way, and there may be a collision between these large disc galaxies, the likely remnant being a bloated amorphous 'star pile' resembling an elliptical galaxy.) Many big galaxies, particularly those at the centres of clusters, may result from such mergers. This process would obviously result in big galaxies having been, on average, fainter in the past.

There are thus two evolutionary effects, both uncertain (and with opposite sign), either of which could be large enough to mask the difference between a universe with $q = 0.1$ ($\Omega = 0.2$) and $q = 0.5$ ($\Omega = 1$). More recent work (based on, for instance, the CDM model) combines these effects – galaxies grow hierarchically, and a new burst of star formation is triggered by each merger event.[58]

The possibility of reaching higher redshifts (though crucial for understanding the astrophysics of galaxies) does not help much in determining q, because the evolutionary effects are then even larger. And the evolutionary effects on quasars etc. are more uncertain still (see Chapter 4). The phenomena of galactic evolution and cosmic expansion are so closely linked that there is little prospect of using galaxies for 'geometrical' cosmological tests until our astrophysical understanding greatly deepens.

There has been a recent revival of interest (and even of optimism) regarding some other probes of the geometry of the universe which bypass the uncertainties of galactic evolution:

(i) Supernovae of type 1 can be detected[59] at $z \gtrsim 0.5$. There is little reason to expect their mean properties to depend on cosmic epoch, so if they are adequate standard candles they offer a possible test of q.

(ii) The radio emission from strong radio sources (the ones first studied by Ryle (see p. 5) comes from lobes containing relativistic plasma and magnetic fields which generally lie symmetrically on either side of the optical galaxy, and whose angular sizes can be readily measured. The very extended sources are undoubtedly influenced by the external medium – this medium itself evolves in an uncertain way, and so introduces the same evolutionary uncertainties as studies of galaxies and clusters. But there may be less reason to expect the typical properties of the *compact* components, lying deep inside active galaxies, to depend on cosmic epoch. Preliminary studies of these[60] tentatively support a value $q = 0.5$ (or $\Omega = 1$).

(iii) The probability that high-z quasars display gravitational lensing depends on the path length (technically, the 'affine distance'), and therefore on the form of $R(t)$. Less than 1 per cent of even the highest-redshift quasars display the multiple images, separated by a few arcseconds, characteristic of lensing by an intervening galaxy. This is consistent with simple estimates based on standard cosmological models. However, if a non-zero cosmological constant Λ allowed the universe to be older than the inverse Hubble constant, the affine distance to high-z quasars would be larger than in 'standard' cosmologies. Intervening galaxies would then act more efficiently as lenses, creating more multiple images, and with wider angular separations, than are observed.[61] There are uncertainties here due to the possible evolution of the lensing galaxies, but this method may be accurate enough to constrain models with large Λ-terms.

Hubble's constant

There is a continuing effort to firm up the cosmic distance scale and determine the Hubble constant H_0, which we have parameterised as $50h_{50}$ km s^{-1} Mpc^{-1}; it is beyond the scope of this book to review these efforts.[62] However, it is fortunate that the problems of H_0 are at least partially disjoint from those involved in measuring q and Ω. When masses are determined from the virial theorem, or from large-scale streaming motions, the inferences about Ω are independent of the value of h_{50}. The inferred contribution to Ω from diffuse gas in clusters of the intergalactic medium tends to be higher for low h (though the precise dependence depends on the physical process involved).

There are two respects in which h_{50} enters sensitively into the topics discussed here.

(i) The age of a universe of given Ω scales as ${h_{50}}^{-1}$. Astrophysical estimates of stellar ages therefore constrain the 'parameter space' of acceptable models more severely if h is high – for instance if h_{50} were closer to 2 than to 1, it would be very hard to sustain support for a standard Einstein–de Sitter model with $\Omega = 1$, for which the time since the big bang is only 2/3 of the Hubble time.

(ii) The processes of primordial nucleosynthesis depend on the baryon density when the universe was cooling through the temperature range 10^{10}–10^{9} K, and therefore simply on the baryon-to-photon ratio η. The consequent upper limit on Ω_b then scales as h^{-2}. If h_{50} were as high as 2, baryons could not contribute more than 2–3 per cent of the critical density.

For both of these reasons, cosmologists will feel uncomfortably constrained if the observational determinations, as they improve, converge towards the upper end of the range of values that are now being advocated.

Clues from the microwave background

Initial fluctuations imprinted in the early universe cannot 'know' what is special about galactic mass. The universe's overall observed homogeneity implies that $(\delta\rho/\rho)$ falls off towards large scales, but fluctuations would nonetheless be expected. Indeed, a natural conjecture, due originally to Harrison and Zel'dovich,[43] might be that the metric (or potential) fluctuations ψ are the same on all scales. If the perturbations had grown unimpeded by pressure, viscosity, etc., but were still in the linear regime, the resultant spectrum of density fluctuations would have the form $(\delta\rho/\rho)$ $M^{-2/3}$ (cf. Figure 14).

An important line of evidence on linear fluctuations, especially those on large scales, comes from the angular distribution of the microwave background. As we look further out beyond the quasars, to redshifts $z \gg 5$, we probe an epoch before any nonlinear structures had condensed out. The microwave background effectively comes from a surface at a redshift $z = 1000$. The imprint of large-scale inhomogeneities that were already present at that epoch would render the background radiation slightly non-uniform over the sky: photons originating in an incipient supercluster straddling the last-scattering surface would suffer an extra gravitational redshift climbing out of the associated potential well, and should therefore appear slightly cooler than those emerging from an incipient void (see Figure 18).

The strongest prediction of inflationary universes is that $\Omega = 1$; however, they lead also to the expectation that the fluctuations would indeed approximate closely to the Harrison–Zel'dovich scale-free spectrum: curvature fluctuations, equivalent to fluctuations in the gravitational potential, should exist with almost the same amplitude on all scales. This implies that there is a cosmic 'magic number' ψ, related to the density perturbation at recombination by a simple formula. The largest gravitational potential wells in the present universe are those of rich clusters, with virial velocities of the order of 1000 kilometres per second. The corresponding dimensionless potential is $(v_{\text{virial}}/c)^2 = 10^{-5}$, and this suggests that, if the metric fluctuations are indeed the same on all scales, ψ has roughly this value.

If metric fluctuations were not already present, at the recombination epoch, with amplitude $\sim 10^{-5}$, then some force *more efficient* than gravity would have been needed to pull together the clusters and superclusters by the present time. There has consequently been a 20-year programme to search for the imprint of these fluctuations in the microwave background. Radiation from an incipient cluster on the surface of last

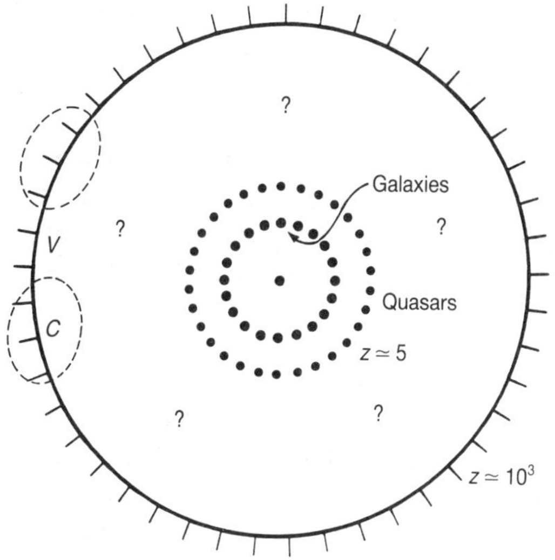

Figure 18 This diagram illustrates various 'redshift shells', looking back along our past light cone, in terms of the Robertson–Walker radial coordinate r. The region $5 \lessapprox z \lessapprox 1000$ corresponds to the era $10^6 \lessapprox t \lessapprox 10^9$ years. Further evidence on this region would help to distinguish between the scenarios for galaxy formation shown in Figure 13. Inhomogeneities in the gravitational potential which straddle the 'last-scattering surface' at $z \approx 10^3$ – protoclusters (C) or protovoids (V) – would create angular structure in the microwave-background temperature, with $\Delta T/T$ of the order of the curvature fluctuation amplitude ψ. (On smaller angular scales there would be a significant contribution to $\Delta T/T$ from Doppler effects; $\Delta T/T$ on small angles may also be attenuated owing to the fact that the universe would have changed only gradually from being opaque to being transparent, so the last scattering surface is not completely sharp.)

scattering would appear slightly cooler, because it has to climb out of an extra potential well. Progressively better upper limits were set by a variety of experiments, and the first positive results came in 1992 from the differential microwave radiometer (DMR) experiment on the COBE satellite.[10] This pinned down the value of ψ relatively unambiguously, to a value lying within the expected range. It offered direct evidence of large-scale fluctuations whose origins must lie in the ultra-early universe (see Chapter 6).

We can be confident of rapid progress in the measurement of microwave-background fluctuations. Already a ground-based experiment in Tenerife,[11] scanning a strip of sky, has found features that correlate with the first-year COBE map. And the second and third years of data should enhance the signal-to-noise so that individual features of the COBE all-sky data (rather than just the excess variance over the sky) become highly significant. Moreover, other experiments are already yielding positive results[12] which can be correlated with these measurements. The present COBE data refer to angular scales of 7 degrees and more, corresponding to present linear scales exceeding $500h_{50}$ Mpc (for $\Omega = 1$). To directly probe the progenitors of clusters and superclusters requires measurements with angular resolutions of a degree or less. The predicted temperature fluctuations on these small scales are sensitive to the cosmological models and to the thermal history of intergalactic gas, and are specially important for these reasons also.[12,63]

The COBE results should be seen as a key step in an ongoing programme of progressively improving techniques involving many groups. The early universe was smooth in the same sense that the surface of the ocean is smooth. There is a well-defined mean curvature, perturbed by waves or ripples. If you look down from the air on an ocean, you may first see just overall smoothness. But as your vision sharpens, you begin to discern some waves. A further modest improvement allows you to study wave statistics in detail. Are the waves Gaussian? How does their amplitude depend on scale? This is a metaphor for the exciting stage we are now entering in the study of the microwave background.

Dissipative effects for the baryon component

The process of gravitational clustering in the expanding universe is dominated by the dark matter. The outcome depends on the initial fluctuation spectrum, and also on the assumed Ω. The mass distribution of isolated virialised systems can in principle be modelled by N-body simulations, starting from specific linear perturbations soon after t_{rec} (cf. Figure 13). But even if the dissipationless clustering of the dark matter is accurately modelled, the fate of the baryonic component – how much gas falls into each potential well and how much is retained – involves complex gas dynamics. To predict what the universe would actually look like – the luminosity function of galaxies and how they are clustered – we need to understand how the baryons behave: they are influenced by many other physical effects apart from gravity. Baryons would settle into virialised halos of dark matter in the mass range 10^8–$10^{12} M_\odot$: for larger masses, dissipative cooling may be inefficient for the reason mentioned earlier (cf. Figure 8); below $10^8 M_\odot$ the potential wells may be too shallow to capture and retain primordial gas.

There are physical reasons for expecting the efficiency of bright galaxy formation to depend sensitively on (for instance) the depth of the halo potential wells. Bright galaxies would be more clustered than the mass[64] for the same reason that in an ocean swell the highest waves come in groups: peaks are more likely to be exceptionally high if they are superimposed on a very-large-scale positive fluctuation (an incipient cluster) rather than in an incipient void (Figure 19).

The degree of 'biasing' is important in estimations of Ω from large-scale streaming (Equation (1)). But biasing is not just an

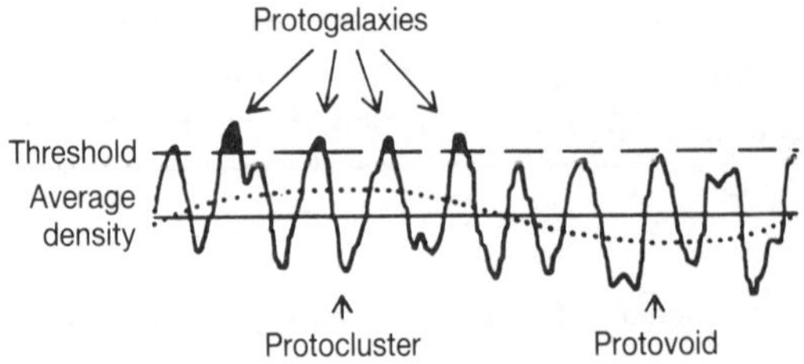

Figure 19 If galaxies form preferentially from exceptionally high-amplitude peaks in a Gaussian density field, they display enhanced clustering (or biasing) because the probability of a high peak is sensitive to whether or not there is a positive contribution to the amplitude from longer-wavelength modes.

ad hoc contrivance which makes it easier to reconcile the theoretically attractive $\Omega = 1$ model with apparently conflicting evidence. There are good reasons for doubting whether bright galaxies would track the overall mass distribution exactly.[65] However, so much complex physics is involved that we cannot realistically expect the efficiency of galaxy formation, in all locations and on all scales, to be well described by a single 'bias parameter' b.

Is any simple hypothesis compatible with all the data?

If Ω is known, along with the amplitude ψ and the nature of the dark matter, the evolution of fluctuations into the nonlinear regime can be followed numerically. The aim is to test whether, for natural assumptions about the initial fluctuations, the present dark-matter distribution is consistent with what is observed and inferred on the scale of galaxies, clusters, and superclusters.

There is a general consensus that a model containing pure 'hot' dark matter – in other words just neutrinos – does not fit the present data.[66] What is more controversial is which of the other models is an adequate fit. The main difficulty in confronting the outcome of these calculations with the real universe is that, until recently, simulations have only really included non-dissipative gravitational matter. They therefore predict the present-day distribution of the *dark* matter. This may be relevant on large scales, where things are still linear, and there has been no chance for segregation of baryons and dark matter to occur. But on smaller scales we can more readily observe the distribution of the *bright galaxies*. And the formation of these huge spinning discs like Andromeda and our Milky Way (and of virialised clusters like Coma) involves a lot more physics.

Theoretical fashions are often transient. But the so-called 'standard CDM model' has been investigated for more than a decade,[42,44–46,67] and throughout that time has served as a benchmark for comparisons with data. This model is specified as follows:

(i) The dominant dark matter is in some non-dissipative 'cold' form (e.g. WIMPs or invisible axions).
(ii) Ω is unity.
(iii) The initial fluctuations are Gaussian, with a scale-independent (Harrison–Zel'dovich) spectrum, the baryon-to-photon ratio η being the same everywhere.
(iv) The biasing of the galaxy distribution relative to the dark matter is described by a single parameter b. (Some degree of 'biasing' may arise naturally because the tightly bound halos (in which bright galaxies would form) are more highly clustered than the matter in general owing to the behaviour of high peaks in a Gaussian random field (cf. Figure 19).)

When the amplitude of the initial fluctuations is appropriately normalised, the standard CDM model accounts very satisfactorily for the properties of galactic halos, groups, and clusters. This success therefore offers circumstantial support for the idea that the dark matter is in WIMPs or axions. As discussed further below, however, the fluctuations on the scales of superclusters, and the value of the Harrison–Zel'dovich amplitude ψ inferred from the microwave-background anisotropies, seem larger relative to those on smaller scales than would be expected in the 'standard CDM model'. If borne out by future data, this would require modification of at least one of the four assumptions above. However, the modifications could involve assumptions (ii)–(iv), and so would not necessarily discredit WIMPs or axions as dark-matter candidates.

The standard CDM model offers a useful template for comparing the results on different scales (see Figure 20 and caption). The amplitude inferred from COBE is roughly consistent with that inferred from large-scale structure and streaming. However, extrapolation of a spectrum with this normalisation down to the scales of clusters, small groups, and individual halos yields an inconsistency.[68] The latter data are well fitted by the *shape* of the CDM spectrum, but only if the amplitude is a factor of 2 lower than is required for a good fit on larger scales. Until a year ago, the small-scale normalisation seemed more reliable. However, we now have better estimates of large-scale structure, as well as the COBE point, and so seem forced to conclude that the standard CDM spectrum has the wrong shape. Something is needed which rises less steeply towards smaller scales than CDM does.

A number of possibilities have been suggested. The initial curvature-fluctuation spectrum could be 'tilted', so that it rises towards larger scales relative to the Harrison–Zel'dovich spectrum.[69] Another acceptable fit comes from a CDM model, modified by the further assumption that the heaviest type of neutrino has a mass in the range 5–10 eV, and thereby contributes 0.3 to Ω. If beam experiments or solar-neutrino observations were to support such a neutrino mass, many cosmologists would be very happy to espouse this so-called 'hybrid' model for the dark matter.[70]

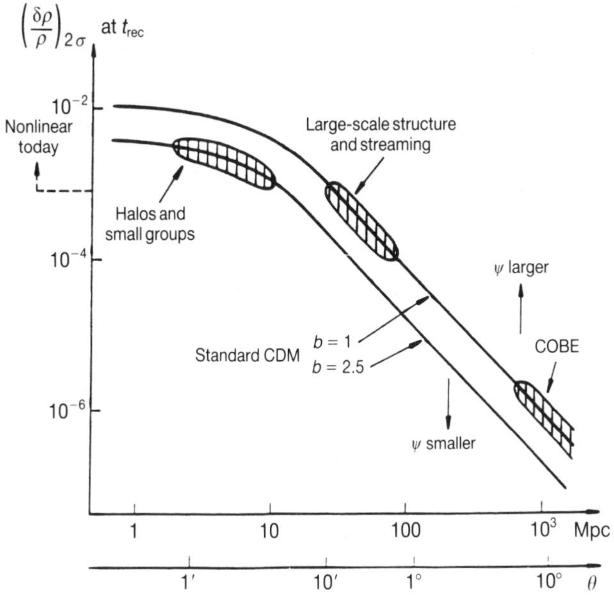

Figure 20 The vertical axis shows the amplitude of '2σ' density fluctuations at the recombination epoch ($z = 1000$). A 'cold-dark-matter' (CDM) model with scale-independent (Harrison–Zel'dovich) initial curvature fluctuations yields the continuous lines. Two different normalisations are plotted, corresponding to particular values of the 'biasing parameter' b. The COBE measurements on angular scales $\gtrsim 10$ degrees, and the evidence on large-scale clustering and galaxy motions, favour the upper of the two curves; but the lower curve gives a better fit to the present-day properties of individual galaxies and small groups. Taken together, the data seem to require a spectrum that has relatively less power on small scales than the CDM model, but it is gratifying that it can all be fitted to within a factor of 2, and that the subject has advanced to the stage when we can discriminate among (and refine) the simple models. The existence of quasars with $z \approx 5$ sets a *lower* limit (not plotted) to $\delta\rho/\rho$ on scales ~ 1 Mpc. The linear scale on the horizontal axis corresponds to a Hubble constant of 50 km s^{-1} Mpc^{-1}($h_{50} = 1$).

The simulations must be matched to the clustering data at the present epoch. However, when this has been done there is a separate test: does the model predict the *observed z-dependence* of the clustering? There is a general trend for the evolution to be more recent when Ω is high. For low Ω, the large-scale structure would be 'frozen-in' for $(1 + z) < \Omega^{-1}$. Clusters of galaxies with z up to unity can now be probed optically, and via their X-ray emission. Indeed, an exciting recent discovery[71] is that the X-ray emission from clusters is increasing with cosmic time, even in a sample restricted to redshifts $z < 0.5$. This is qualitatively consistent with the expectations of hierarchical models, according to which big clusters would have formed only recently as the outcome of mergers of smaller ones.

But it is the *quasars* which provide the most direct clues to the early stages of galaxy evolution, and to what happened soon after the first bound structures collapsed. We turn to them in the next chapter.

4

Quasars and their demography

Quasars and the epoch of galaxy formation

The z-dependence of the quasar population

Galaxies – disc-like spiral systems like our own Milky Way, together with the less photogenic ellipticals – are the basic building blocks of the cosmos. The observed light from most galaxies is essentially the total output from their billions of constituent stars. It has been known for more than 30 years that some galaxies are more than just assemblages of stars and gas: some have a bright central 'nucleus', whose emission does not come from normal stars. The most extreme 'active galactic nuclei' (AGNs), where a central object no bigger than our Solar System vastly outshines all the 10^{11} stars in its host galaxy, are the *quasars*.

How and why AGNs form and evolve is still, in many respects, mysterious. There is a strong consensus that the power output derives primarily from gravitation (rather than, for instance, nuclear energy) and that, to yield adequate efficiency, a relativistically deep potential well must be involved. The detailed modelling of the primary power output is complex, and in many respects still controversial. But it seems that, in the centres of some galaxies, stars and gas become so close-packed that some kind of runaway catastrophe is triggered. Violent activity may be a relatively short-lived phase in the life of a galaxy, but the energy output during a quasar's active phase may be equivalent to the rest mass of $>10^7$ suns.

Quasars are cosmologically important for two reasons.

Firstly, they tell us that, even at high redshifts, galaxy formation had proceeded far enough to allow such objects to form. Secondly (as described in Chapter 5), they serve as probes for the intervening medium along the line of sight.

Several thousand quasars have now been catalogued.[72] Systematic surveys allow astronomers to deduce how many quasars there were at different cosmic epochs. The most remarkable feature of the quasar population is that it declines sharply between $z = 2$ and the present epoch ($z = 0$). This has been known since the 1960s, and was prefigured even earlier by the (then-controversial) radio-source counts.[5] Our nearest bright quasar, 3C 273, is 500 Mpc away. At $z = 2$–2.5 the nearest quasar would have been 30 times closer, and as bright as a 4th-magnitude star. (It is an anti-anthropic irony that the best time to be an astronomer was at that early era, before the Earth had formed.) We might, of course, have expected quasars to be closer in past times (by a factor $(1 + z)$), because the entire universe was then denser overall. But the effect is much larger than can be accounted for this way: at $z = 2$, quasars were a thousand times more common, relative to galaxies, than they are today.[73,74]

Quasar activity seems to have peaked sharply at the epoch when the universe had about 1/3 its present scale. It has proved much harder to discover quasars with still-larger redshifts – until 1987, none were known with redshifts exceeding 4, though there are now more than 50 in this category, the most distant[6] being PC 1247 + 3406, with $z = 4.89$ (Figure 3). The difficulty of reaching higher redshifts is only partly the effect of greater distance: the quasar population genuinely thins out[74] (see Figure 21). The comoving density seems to decrease at $z > 3.5$, though the steepness of this trend is still controversial (see Figure 22).

In an Einstein–de Sitter cosmology, time and redshift are related by

$$t(z) = 13.1 h_{50}{}^{-1}(1 + z)^{-3/2} \text{ years.} \tag{2}$$

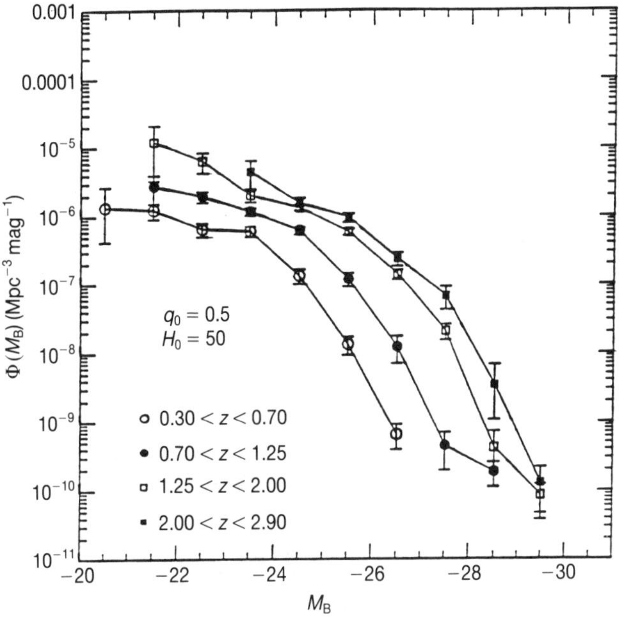

Figure 21 The luminosity function for quasars in different redshift ranges, showing the much higher comoving density of luminous objects at redshifts 2–2.5 than at recent epochs. (From Boyle, B. J. *et. al.* 1990, *Mon. Not. Roy. Astron. Soc.* **243**, 1.)

The quasars with highest-known z formed when (according to Equation (2)) the universe had been expanding for 10^9 years and was only 7 per cent of its present age. By that time, some galaxies must have formed and already evolved to the stage where runaway activity gets triggered in their nuclei. The appearance of quasars so early in cosmic history is an important constraint on galaxy-formation scenarios – in particular, it is an obvious embarrassment to scenarios (e.g. the 'top–down' neutrino-dominated model) which do not predict that the action would start as early as this.

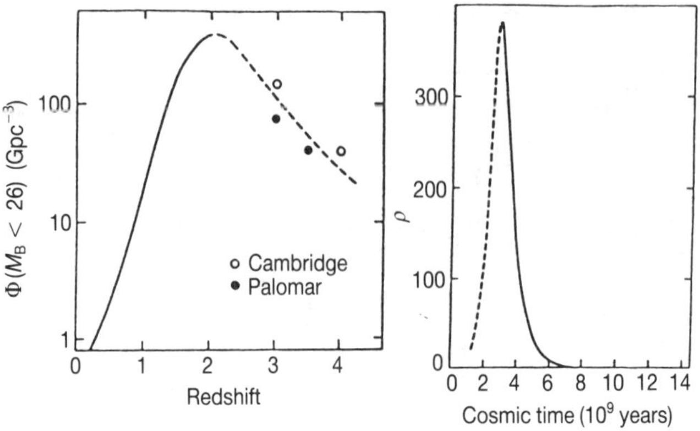

Figure 22 The comoving density of powerful quasars is given on the left as a function of redshift z. (Beyond $z \approx 3$ there is a falloff, but the quantitative details are uncertain. The filled and open circles correspond to the results from two different surveys.) The same comoving densities are, following Schmidt, replotted on the right on a linear scale; the horizontal axis in this plot is cosmic time, assuming an Einstein–de Sitter cosmology with $h_{50} = 1$. This plot dramatises the relative brevity of the 'quasar era', when the universe was 2 to 3 billion years old. (From Schmidt, M. 1989, *Highlights of Astronomy* **8**, 31.)

Theoretical estimates of the redshift of galaxy formation

In the 1970s, it was widely believed that 'galaxy formation' happened at earlier epochs than were directly observable. Since that time, not only have quasars been detected at larger redshifts, but theorists' estimates of the redshift of galaxy formation have come down. This is because galaxies (with their dark halos) are more extended and diffuse than was previously suspected. Current ideas on extended halos, and on the origin of galactic rotation, suggest – almost irrespective of the cosmogonic model – that galaxy formation, however early it may have started, must still have been going on at eras that we can directly observe (corresponding to redshifts $z < 5$).

At the epoch corresponding to $z = 5$, the cosmic-expansion timescale (Equation (2)) is long compared with the dynamical timescale within the 'luminous' part of a typical galaxy. But it is *not* long compared with the timescales for *extended halos*. In a galaxy like our own, the freefall timescale from a radius of 100 kpc is 10^9 years; the outer part of a galactic halo could therefore not have recollapsed and virialised until the universe was 2 billion years old – in other words, not much sooner than the epoch corresponding to $z = 2$. (Indeed, more realistic models, where halos build up hierarchically from highly non-spherical perturbations, suggest that formation could be more recent still.)

A further argument suggests that not only the outer halos, but even the *discs* of galaxies like our own (with radii ~ 10 kpc), are relatively recent acquisitions. This argument[75] is based on considering where the angular momentum of galaxies came from. Angular momentum cannot have been effectively 'stored' in the early dense stages of the universe. Protogalaxies must have acquired their spin by tidal interactions with their neighbours. This process depends on the initial perturbations having been highly non-spherical (as expected if they were part of a broad spectrum of Gaussian fluctuations), so that the protogalaxies typically have large quadrupole moments. 'Tidal torquing' between protogalaxies would be maximally efficient when they are just turning around – at earlier times, they are merely small fractional enhancements in the mean density; at later times, when the density contrast is very large, their separations are much larger than their characteristic sizes, so the tidal effects are weak compared to the basic inverse-square attraction. An often-used dimensionless measure of a system's angular momentum is $\lambda = G^{-1}JE^{1/2}M^{-5/2}$, where J is the angular momentum and E the gravitational binding energy. For a protogalaxy, λ is roughly the ratio of the rotational

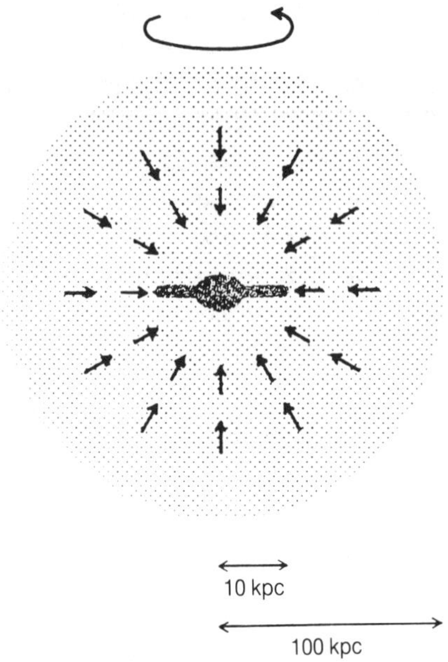

Figure 23 Tidal torques typically impart less than 10 per cent of the rotational velocity needed for centrifugal support. The material that ends up in a disc of radius 10 kpc must therefore have fallen in from $\gtrsim 100$ kpc. This infall timescale is $\approx 10^9$ years, so the formation of galactic discs like those in our Milky Way and in Andromeda cannot have been completed before the epoch corresponding to a redshift of 2 or 3.

velocity to the virial velocity at the moment of turnaround. Both analytic and numerical estimates indicate that tidal torques would have given typical protogalaxies values of λ in the range 0.03–0.06. Consequently, the material in a protogalaxy cannot achieve rotational support unless it contracts dissipationally by a substantial factor (see Figure 23). For a rotationally supported disc of ~ 10 kpc to form, baryons must have acquired their angular momentum at ~ 100 kpc. (This assumes a heavy halo extending out to ~ 100 kpc, and also that the baryonic matter does not transfer any of its angular momentum to the halo during infall – more realistic models may require a still larger 'lever arm', implying even slower and more-recent formation of galactic discs).

The assembly of large galaxies (even though, in some 'bottom–up' scenarios, it could have started very early) could therefore not have been completed before the epoch ($z = 2$–2.5) when the quasar population peaked. So quasar formation is intimately linked to the accumulation process that forms the galaxies themselves. High-redshift quasars therefore serve as probes of a cosmogonically important era when conditions should certainly have been interestingly different from the present time.

In outlining some current ideas, I shall start off with inferences that follow directly from observations, and conclude with a brief (more speculative) summary of how quasars can be related to the formation of galaxies discussed in the previous chapters.

How many quasars have there been?

Quasars are relatively rare cosmic beasts. At the present epoch there is less than one for every 10^5 galaxies; even during the 'quasar era', at redshifts of 2–2.5, quasars were about 100 times less common than normal galaxies.[73,76] At first sight, one might infer from this that only one galaxy in 100 has ever indulged in quasar activity, and that we should not expect to find remnants in more than about one per cent of galaxies today. But there is an alternative possibility. Figure 22 depicts the rise and decline of a *population*: the curve need not represent the life-cycle of a typical object. Just as in a biological population, many short-lived generations could be born, evolve, and die over the period where the overall population rises and falls. If there were many generations of quasars, each living much less than 10^9 years, then we would expect more remnants; the mass of each remnant, being related to the total energy expended over a quasar's life, need not then be so colossal.

The integrated background light from quasars, estimated from quasar surveys, amounts to an energy density of around $3000M_{\odot}c^2$ per cubic megaparsec. This output is dominated by objects with redshift $z \approx 2$, with the luminosity function shown in Figure 21. The present spatial density of galaxies is known, and we can therefore estimate the mean amount of energy per galaxy associated with quasar activity. The mass-equivalent of this energy amounts to around $10^7 h_{50}{}^{-3} M_{\odot}$ per bright galaxy.[77]

Many features of quasars remain enigmatic: active galactic nuclei display many phenomena on various scales and different wavebands, and it is hard to fit them into a single pattern. But there is less room for doubt about what a *dead* quasar should be: there seems no way of evading the conclusion that a substantial fraction of the mass involved must eventually collapse to a massive black hole. If we optimistically assign an efficiency of 10 per cent to the overall energy generation in quasars, we must then conclude that their black-hole remnants amount to an *average* of about $10^8 h_{50}{}^{-3} M_{\odot}$ per bright galaxy.

But is this mass 'shared out' among all galaxies (implying that there would have been many generations of short-lived quasars) or do quasars form, and persist for $> 10^9$ years, in a small favoured fraction of galaxies? It seems unlikely that the most luminous quasars could last as long as 10^9 years: the resultant masses would then be $10^{11} M_{\odot}$ unless their emission were narrowly beamed towards us. However, the situation is less clear for lower-powered objects. We do not know whether individual quasars follow a standard evolutionary track, with bright and faint phases. The lack of direct evidence on individual quasar lifetimes, or the 'duty cycle' for the different kinds of activity in galactic nuclei, is a severe stumbling block to our attempts to understand the evolution depicted in Figure 22.

Quasar masses and efficiencies

There are in principle other ways of estimating quasar masses. The optical spectra are dominated by emission lines from fast-moving clouds of gas, heated by continuum radiation from the central object. The distance of the clouds from the centre can be estimated from the ionisation state of the gas (it is typically a few light-weeks), and their velocities from the line widths (they are several thousand km s^{-1}). If the clouds have gravitationally induced motions – if they are, for instance, in freefall or in orbit – one can therefore infer the central mass; if the gas is flowing out in a wind, this argument yields merely an upper limit.

A second argument depends simply on noting that, if quasars are powered by accretion, then outward radiation pressure cannot overwhelm gravitation, implying (apart from beaming corrections) a lower limit to the mass. If electron scattering produces the main opacity, this limit is $7 \times 10^7 (L/10^{46}$ erg $s^{-1})M_\odot$. Zel'dovich and Novikov[78] advanced this argument right back in 1964, and there is little cause to modify this conclusion of their pioneering work. The masses implied are at least $10^8 M_\odot$ for typical quasars, but $10^9 M_\odot$ for the most powerful ones.

Table 1 contrasts two hypotheses: (i) that there was, in effect, only one generation of quasars, which were long-lived and massive; versus (ii) that there were 50 generations of quasars, so that their individual masses (for a given efficiency) need not have built up to such high values, and quasar remnants would be more numerous. In reality, t_0 will not be a single number, but there would be a spread of ages, perhaps correlated with luminosity L.

There is a characteristic timescale, derivable from fundamental constants, which is the time an object takes to convert its entire rest mass into radiation if it radiates at the Eddington luminosity. This timescale, first explicitly given by Salpeter,[79] is

Table 1

(i)	(ii)
$t_0 \approx t_{Evo}$	$t_0 \approx 4 \times 10^7$ years $\approx 0.02 t_{Evo}$
$M = 2.5 \times 10^9 L_{46} \varepsilon_{0.1}{}^{-1} M_\odot$	$M = 5 \times 10^7 L_{46} \varepsilon_{46} \varepsilon_{0.1}{}^{-1} M_\odot$
$L \ll L_{Ed}$	$L \approx L_{Ed} \varepsilon_{46} \varepsilon_{0.1}{}^{-1}$
Broad-line regions gravitationally bound	Broad-line region *not* gravitationally bound
Very massive remnants in ~2% of galaxies	~$10^8 M_\odot$ remnants in most bright galaxies

$$t_{Sal} = \sigma_T c / 4\pi G m_{proton} = 4 \times 10^8 \text{ years.} \tag{3}$$

Astrophysical studies of accretion indicate that the efficiency ε with which the rest mass of infalling material is converted into radiation is unlikely to exceed 0.1. The characteristic doubling-time for the hole's mass would be εt_{Sal}. It is essentially for this reason that many theorists favour around 50 million years ($0.1 t_{Sal}$) as the 'best bet' for a typical quasar lifetime.

These arguments cannot distinguish between a single active phase and repeated short outbursts whose total duration adds up to the same number of years. (If quasars lived for much less than 50 million years, they would be converting mass inefficiently into radiation unless they had such low masses that gravity could not compete with radiation pressure.)

The figures in the right-hand column of Table 1 suggest that there may be as many black holes as big galaxies. Moreover, by focussing only on powerful quasars, we may be underestimating the mass and number of black holes in galaxies: the postulated 10 per cent efficiency may be over-optimistic for the average AGN, and black holes could in principle form via other routes (or even be primeval). It seems well worth exploring how best they can be detected.

Dead quasars: massive black holes in nearby galaxies

Stellar 'cusps'

Most galaxies which were ever active were active in the remote past. So 'dead quasars' – massive black holes now starved of fuel, and therefore quiescent – should lie about us. A massive black hole manifests itself as a quasar only so long as it is fuelled by capturing gas from its surroundings. But its remnant mass would still, after its 'death', exert a gravitational influence. Stars would tend to be pulled towards it, so that there would be a concentrated spike or cusp in the stellar light distribution around it, owing to fast-moving stars.

The hole's range of influence is determined by how close a star has to get before its orbit 'feels' the hole's gravity, rather than just the integrated gravitational field of all the other stars (and dark matter) in the galaxy. The 'sphere of influence', of radius r_h, is the volume within which the escape velocity from the hole's own gravity (or, almost equivalently, the orbital speed of stars gravitationally bound to it) exceeds the typical velocity dispersion V_c throughout the galaxy. r_h is proportional to the hole's mass M_h and depends inversely on the square of V_c. It is millions of times larger than the 'size' of the hole itself, $r_g = GM_h/c^2$ (see Figures 24 and 25), but still only large enough to subtend an angle of a few arcseconds, even in the nearest galaxies.

There would therefore be two tell-tale signs of a quiescent black hole. The first is simply a central 'blip' in the light profile, resulting from the enhanced concentration of stars within the hole's sphere of influence. This has been searched for; but there is always an ambiguity in that there may be some extra source of non-stellar light at the centre. A second, less ambiguous, signature would be spectral evidence that the stars nearest the centre were moving anomalously fast.

At the distances of even the nearest galaxies, the gravitational effect of the central black hole would be discernible only within

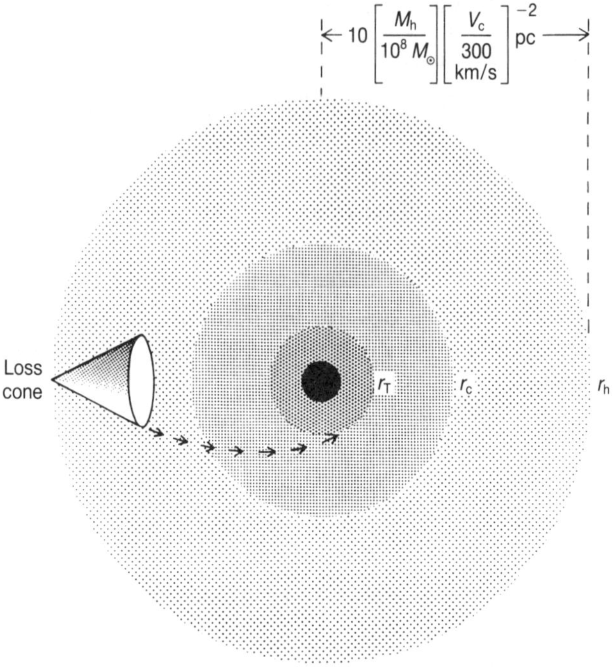

Figure 24 This diagram depicts, not to scale, various characteristic radii around a massive black hole of mass M_h in a stellar system. If the velocity dispersion in the core of the galaxy is V_c, the hole influences the stellar motions within a radius $r_h \approx (GM_h)V_c^2$. Within r_c, stars would be moving so fast that they would be more likely to experience (generally disruptive) physical collisions with each other than to undergo two-body encounters of the kind that can be treated by point-mass approximations. r_c is the radius where the escape velocity from the hole is comparable with the escape velocity $v \approx (2Gm_*/r_*)^{1/2}$ from the surface of a star. Tidal disruptions occur only within a much smaller radius $r_T \approx (M_h/m_*)^{1/3}r_*$. To be disrupted, a star must cross the sphere at $r \approx r_h$ on a nearly radial 'loss-cone' orbit.

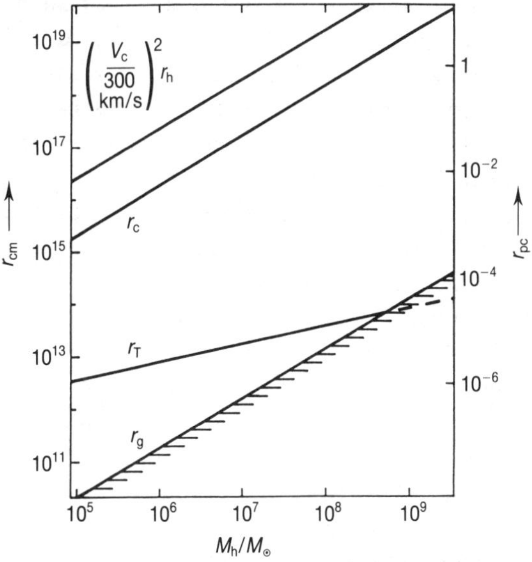

Figure 25 The various radii depicted in Figure 24 are plotted here, on a logarithmic scale, as a function of the hole mass M_h. The hole's gravitational radius is $r_g \approx 1.5 \times 10^{13}(M_h/10^8 M_\odot)$ cm. The radii r_c and r_T are plotted for a solar-type star with $v_* \approx 1000\ \mathrm{km\ s^{-1}}$. Note that a hole of $\gg 10^8 M_\odot$ can swallow solar-type stars (though not, of course, giants) without first disrupting them. When $r_T \approx r_g$, tidal disruption effects would be restricted to the general-relativistic domain where the hole's tidal effects cannot be adequately modelled by an r^{-3} Newtonian approximation.

an angular distance of a few arcseconds of the galactic nucleus (corresponding to r_h in Figure 24). The first indications of such effects, dating from the late 1970s, related to the giant elliptical M87, where a mass of 3–5 × $10^9 M_\odot$ was claimed on two separate grounds:[80] a central peak in the light profile was attributed to the enhanced concentration of stars around a central dark mass; and the velocity dispersion appeared to rise from 300 km s^{-1} to $\gtrsim$ 500 km s^{-1}, as would be expected if there were a central '$1/r$' potential. This discovery stimulated theoretical attempts to evade this conclusion, especially by invoking anisotropic velocity distributions. More recent observational work suggests that the situation is less straightforward. Much of the central peak in the light is now believed to be non-stellar in origin – this is not unexpected, since M87 is a radio source, whose well-known jet testifies to continuing non-thermal activity, at some level, in its nucleus.

Optical astronomers have now found evidence for dark central masses in several nearby galaxies. These galaxies, unlike M87, do not currently seem 'active' even at low levels; and because they are more than 10 times closer, central-mass concentrations might reveal themselves even if they were 'only' 10^6–$10^7 M_\odot$. The most interesting and convincing case involves our nearest large neighbour in space, the Andromeda galaxy, M31. In the 1960s, Schwarzschild and his Princeton colleagues[81] had flown a small telescope, 'Stratoscope', in a high-altitude balloon to avoid the blurring effect of the lower atmosphere. They discovered that the stars in the central few light-years of M31's core had a flattened distribution. In the last few years, spectra have revealed that the stellar velocities rise towards the centre.[82] Also, the flattened stellar system does indeed seem to be rotating around a concentrated central mass of around $3 \times 10^7 M_\odot$.

M31 is not the only nearby galaxy for which this type of evidence exists. The small galaxy M32 apparently[83] harbours a central dark object of $\sim 5 \times 10^6 M_\odot$. Another interesting case is the Sombrero galaxy.[84] This has, like Andromeda,[82] an apparently rotating stellar core, with a central mass of almost $10^9 M_\odot$. The best ground-based measurements, from the CFHT (Canadian-French Hawaii Telescope) 4-metre telescope in Hawaii,[85] achieve a resolution as sharp as 0.3 arcseconds. But still better data from the repaired Hubble Space Telescope should soon firm up the evidence in all these systems.[86,87]

Even if we are convinced that these galaxies contain a concentrated dark mass at their centres, does this have to be a black hole?[87] There is no central spike in the light distribution associated with the putative gravitating mass in the centre of M31, implying that, whatever it may be, its 'mass-to-light ratio' is at least 35 times the Sun's. Could there be an unusual concentration of faint stars near the centre? A cluster 10 light-years across (about 1 arcsecond at M31's distance), though unappealingly '*ad hoc*', cannot be completely excluded. But it could be ruled out by the sharper images that the Hubble Space Telescope can provide: if the stars closest to the centre were moving still faster, implying that the dark mass was concentrated within the central 0.1 arcseconds, not merely the central 1 arcsecond, this would rule out a compact cluster of dark stars, because such a compact cluster would evolve quickly, owing to stellar encounters; it would therefore be implausible to find it in a 10-billion-year-old galaxy.

Flares from tidally disrupted stars?

These massive black holes are not unexpected – indeed if we make the best guess about quasar lifetimes (and how many generations have lived and died) we would not be surprised to find a black hole in most galaxies (cf. Table 1). But before accepting this conclusion we should address a further worry: can such a big black hole really lurk in these galaxies without showing other evidence of its presence? We are used to the idea that black holes (when accreting) are ultra-efficient radiators. But there is no sign of such activity in M31: the upper limit is no more than a ten-thousandth of a quasar. So could a black hole be so completely starved of fuel that it does not reveal its presence as at least a 'miniquasar'?

There is no a priori reason why the centre of M31 could not be 'swept clean' of gas. The *star* density, however, is known to be high – after all, if the stars were not closely packed near the centre of the galaxy we would not have evidence for the black hole at all. Each star traces out a complicated orbit under the combined influence of all the other stars and of the hole itself. The orbits gradually change, or 'diffuse', owing to the cumulative effect of encounters with other stars. There is a chance that these encounters will shift a star onto a nearly radial orbit which brings it very close to the hole; the occasional star may even plunge right into the hole.

There is a limit to how closely a star can approach a black hole without suffering damage. A very compact star (a white dwarf, for instance) could fall inside a massive black hole more or less intact. Larger stars, however, are more sensitive to tidal effects. One can calculate how close a passing star can get before tidal forces tear it apart. The radius r_g of a black hole scales with its mass, whereas the 'tidal radius' r_T goes only as the cube root (see Figures 24 and 25). A hole as large as the one claimed to exist in M87 could swallow a solar-type star without disrupting it. However, if the hole's mass is 10^6–$10^8 M_\odot$ (the range relevant to the nearest galaxies) then, for stars like the Sun, the tidal radius is 10–100 times larger than the hole itself.

It is an intricate (though tractable) problem in stellar dynamics to calculate the chance that a star passes within the central hole's tidal radius. Such events happen about once every few thousand years – the exact rate depends on the statistics of the stellar orbits, and particularly on how quickly the radial 'loss-cone' orbits are replenished.

The flare resulting from a disrupted star could be the clearest diagnostic of a black hole's presence.[88] This phenomenon poses an as-yet-unmet challenge to computer simulations. The relatively crude calculations so far carried out should, nevertheless, convey the essence of what goes on. About half the debris would fan out, with speeds up to 10^4 km s^{-1}, on hyperbolic orbits; the other half would become bound to the hole and would gradually swirl down into it.[88,89] The nucleus would flare up to a brightness much greater than a supernova – almost indeed to a quasar level – but only for about a year. It is hard to calculate how much of this radiation emerges in the visible band, rather than the UV (or even X-ray) parts of the electromagnetic spectrum.

A second question is how fast the brightness fades: how long does it take for the last dregs to be digested? This is important because we want to know whether the luminosity would fade below detectable levels before (10^3–10^4 years later) the next stellar disruption occurs. The details are still controversial – radiative processes and gas flows tend to be more difficult to calculate than dynamical phenomena, where stars can be treated as Newtonian point masses (just as, in Chapter 3, we noted that the hardest aspects of galaxy formation to quantify are those involving the baryonic component).

The predicted flares may be the most robust test of the black-hole hypothesis. Clearly we should not expect one from M31 during our lifetime. But if most galaxies harboured black holes, then a survey of the nearest 10^4 galaxies should catch a few at the peak of a 'flare', and probably rather more in a state where the effects of the most recent tidal disruption were still discernible. Monitoring programmes should either soon reveal instances of this new phenomenon, or else start to put significant limits on central black holes and/or the stellar distributions around them.

A note on our Galactic centre

It may well be that the largest holes are found in elliptical galaxies. However, the indications from M31 and other disc galaxies suggest that our own Galaxy would be 'underendowed' if there was no black hole at its own centre. There has for many years been dynamical evidence for a concentrated central mass of $3 \times 10^6 M_\odot$. This evidence comes, however, primarily from the motions of gas streams rather than stars. The gas is subject to non-gravitational forces, and may not follow ballistic trajectories; there is therefore some ambiguity. A unique compact radio source appears to lie, almost at rest, at the dynamical centre of our Galaxy. This source can be naturally, though not uniquely, interpreted as an effect of very-low-level accretion onto a massive black hole.

In a review article seven years ago, Genzel and Townes[90] discussed the complex phenomenology of stars and gas near the Galactic centre, and rated the case for a massive black hole as 'substantial but not fully convincing'. The same authors have just produced a new article,[91] and I turned eagerly to see how they had updated their assessment. Their latest verdict is that the case is 'holding up, and perhaps slowly growing, but by no means fully convincing'.

Binary black holes?

There may be central black holes in most galaxies. It also follows from the data in Figures 21 and 22 that most of these had already formed by the time the universe was 2 or 3 billion years old. According to 'hierarchical' cosmogonies, most galaxies would have experienced a merger since that time – indeed, merging galaxies are quite commonly observed. When two galaxies merge, the orbits of their constituent stars will become mixed up, the resultant 'star pile' resembling an elliptical galaxy. If each of the original galaxies contained a massive black hole, these would spiral together into the centre of the merged galaxy, forming a *massive binary*.

The holes eventually approach close enough that each is influenced more by the gravitational pull of the other than by the collective effect of the surrounding stars. The black-hole binary continues to shrink, as it transfers kinetic energy to stars (and perhaps also suffers drag on gas), until it gets sufficiently close for gravitational radiation, operating on a timescale proportional to the fourth power of separation, to bring about eventual coalescence. A powerful burst of gravitational radiation, carrying up to 10 per cent of the mass-energy of the holes, is emitted during the last stages of the inward spiral towards coalescence. These gravitational waves, in the very-low-frequency range 10^{-3}–10^{-4} Hz, could be readily detected, even from high redshifts, by planned space interferometers. However, the discouraging news for experimenters is that the event rate would be less than one per decade, unless there were an extra population of moderate-mass (10^4–$10^6 M_\odot$) holes in addition to those associated with quasars.[92]

It is not completely obvious, incidentally, that the hole resulting from a merger will remain lodged in the centre of its host galaxy. There may be a recoil due to emission of net linear momentum by gravitational waves in the final coalescence.[93] This is a strong-field gravitational effect which depends essentially on there being a lack of symmetry in the system. It can therefore only be properly calculated when fully 3-dimensional general-relativistic calculations are feasible. The velocities arising from these processes, on the basis of very crude approximations, are likely to be several hundred km s^{-1}. If a third hole drifts in before the binary has merged, a Newtonian 'sling shot' may lead to ejection at much higher speeds. So some massive black holes could be hurtling through intergalactic space!

Cosmogonic interpretations of quasar evolution – some speculations

Quasar activity presumably cannot start until some galaxies (or at least their inner regions) have condensed from the expanding universe, so that runaway activity can occur within a localised potential well. The fact that quasars formed so early in cosmic history is an important constraint on models for galaxy formation, particularly on 'top–down' models in which large-scale structures develop before individual galaxies; for instance, the simple adiabatic ('pancake') model dominated by neutrinos cannot, when the amplitude is normalised to fit clustering data or the microwave-background anisotropies revealed by COBE, account for collapsing systems at such high redshifts.

In hierarchical models for structure formation, the earliest quasars would be expected to develop in the first sufficiently massive and deep potential wells that virialise, after the collapse of high-amplitude peaks in the initial dark-matter distribution. In a standard CDM model, the ($>3\sigma$) peaks that collapse at $z = 5$ have masses 10^{11}–$10^{12} M_\odot$, and develop velocity dispersions $V_c = 400$ km s^{-1}. These bound halos should certainly contain sufficient mass to build a quasar and potential wells deep enough to retain baryons efficiently. The main uncertainty is whether (and how quickly) the inner $10^9 M_\odot$ of gas (turning around at a radius of 5 kpc) can lose angular momentum and settle into the centre.[94]

Unless it all converted immediately into stars, the inner $10^9 M_\odot$ of gas would fall inward, on a timescale $\sim 10^8$ years, until angular momentum became important. For a typical value of the λ-parameter in the protogalaxy, rotational support would halt collapse at 100 pc if the gas retained all its initial angular momentum. Star formation could occur during this infall – indeed, this is the way the inner bulge of a large galaxy might form. Figure 26 shows schematically some of the processes involved. Production of heavy elements via high-mass stars can occur during this infall. The 'branching ratio' between star formation and direct infall cannot be predicted.

The dynamical timescale at 100 pc is 3×10^6 years. The next question is therefore whether material rotationally supported at this radius – unprocessed gas, together with material that has been processed through stars – can contract still further towards the centre. Specifically, it is important to know whether this gas can lose its angular momentum in less than 100 orbital periods (i.e. 3×10^8 years). If it cannot, then it will not trigger a quasar as early as the epoch $z = 5$.

The efficiency of angular-momentum transfer is hard to quantify. However, simulations suggest that, in a self-

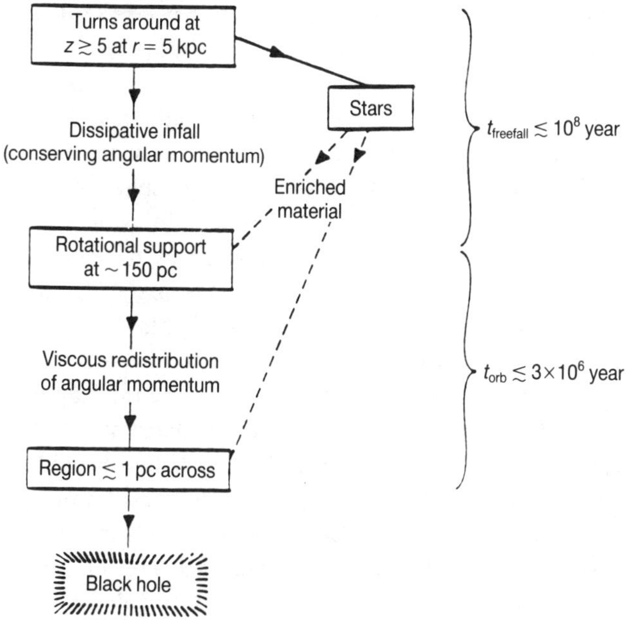

Figure 26 Processes and timescales for the central $10^9 M_\odot$ of baryons in a protogalaxy which starts to collapse at $z \lesssim 5$. This material forms the stellar bulge, and probably also a central black hole. High-z quasars may be manifestations of this process.

gravitating system, it takes only a few dynamical timescales to transport angular momentum outwards.[95] If this is indeed so, 10^8–$10^9 M_\odot$ of baryons (already enriched with heavy elements) could accumulate in a central region less than a few parsecs in size within 10^8 years of the initial collapse. Indeed, such an outcome seems almost inescapable. The only chain of events that would preclude the accumulation (and runaway collapse) of a central mass concentration would be the quick and complete conversion of infalling gas (see Figure 26) into stars that are all of such low mass that they neither expel nor recycle any of their material within 3×10^8 years.

Cosmological N-body simulations show that large halos at the present epoch have a variety of histories – some would have nucleated around a single peak; others may result from a relatively recent merger of systems that coagulated around separate high peaks in the initial density distribution. But in general they form from the inside outwards: the inner $10^{10}M_{\odot}$ may virialise at $z > 8$ even though the halo is not fully assembled until $z < 2$. The baryons most likely to aggregate into a central compact object are precisely those associated with this inner material: it is by no means improbable that a large fraction of *these* baryons ($10^9 M_{\odot}$) would participate in the AGN. A quasar could therefore switch on before the halo was assembled, its active phase being perhaps concurrent with the formation of the stars in the 'bulge'. The host systems would by now have accumulated $> 10^{12}M_{\odot}$ halos. The most promising nearby sites for dead quasars are therefore the centres of very big galaxies, but this does not mean that the onset of quasar activity had to await the assembly of the entire halo.[94]

At first sight, one might think that a hierarchical cosmogony, where the potential wells get progressively deeper and more massive, would predict even bigger black holes, and hence even more powerful AGNs, at recent epochs. But the luminosity function shown in Figure 21 suggests, contrariwise, that *higher-z* quasars typically involve *bigger* black holes.

How could this arise? Black-hole formation requires a deep potential well, but may be more efficient when collapse occurs at high redshifts. This is because systems collapsing earlier tend to have higher density (and therefore allow more efficient dissipation) as well as less angular momentum. On the assumption that each hole maintains an Eddington luminosity for $0.1t_{\mathrm{Sal}} = 4 \times 10^7$ years before fading, specific models along these lines yield gratifying fits to the quasar luminosity function, and to the way it rises and falls between $z = 5$ and the present epoch (Figures 21–22).

The rare *low-z* quasars and powerful AGNs are not in newly formed galaxies: some special environmental influence – for instance, close interaction or merger with a neighbour – must reactivate a small proportion of already-formed galaxies. (It is often conjectured that *all* quasars are triggered by such an interaction. However, at high redshifts no galaxy would have settled down to a stationary axisymmetric structure. The quasar phase is a concomitant of this settling-down process.)

Hubble's great book, *The Realm of the Nebulae*,[96] concludes with these words. 'With increasing distance our knowledge fades and fades rapidly. Eventually we reach the dim boundary, the utmost limits of our telescope. There we measure shadows, and we search among ghostly errors of measurement for landmarks that are scarcely more substantial. The search will continue. Not until the empirical resources are exhausted need we pass on to the dreamy realm of speculation.' This search is continuing, as more powerful telescopes and detectors are being deployed. Ideas on how quasars form are still rather qualitative, and can only be validated (or refuted) when we have better observations of galaxies at high redshift, and of the 'hosts' of high-z quasars, rather than just of the central activity. The empirical data are advancing rapidly. And theorists are injecting a range of not-necessarily-compatible ideas whose vector sum at least pushes our understanding forward. The nature of the activity in the centres of galaxies is still somewhat mysterious, but the key questions are at least in clearer focus.

In 1975 Chandrasekhar wrote,[97] 'In my entire scientific life . . . the most shattering experience has been the realisation that an exact solution of Einstein's equations of general relativity, discovered by the New Zealand mathematician Roy Kerr, provides the *absolutely exact representation* of untold numbers of massive black holes that populate the Universe.' It is now even clearer that there may indeed be as many massive black holes as there are galaxies. Efforts to understand the physical processes associated with them – accretion, tidal disruption of stars, gravitational waves, etc. – may, as a bonus, reveal some distinctive relativistic effects which allow us to test Einstein's theory in the strong-field regime.

5

Some probes and relics of the high-redshift universe

Quasars as probes of intervening gas

A uniform intergalactic medium?

In the spectra of quasars with $z > 2$, radiation emitted blueward of Lyman α is detected, shifted into the visible band (e.g. Figure 3). This simple observation has important implications for the intergalactic medium (IGM), which have been recognised since 1965 and associated with the names of Gunn and Peterson.[98] So high is the cross-section for absorption in the Lyman-α resonance line that no such radiation would get through to us unless the neutral-hydrogen (HI) density were below about 10^{-11} cm^{-3}. It would be astonishing if galaxy formation had 'swept' intergalactic space so completely that the gas density was actually this low. A much more likely explanation is that the universe is pervaded by ultraviolet (UV) radiation intense enough to maintain any diffuse intergalactic medium almost completely ionised, like the interior of an 'HII' region around hot stars. The UV background from quasars, and perhaps also from populations of hot stars in young galaxies, should be intense enough to do this, provided that the total contribution of the intergalactic medium to Ω is not substantially more than the present Ω_b contributed by galaxies and clusters.

The calculated temperature of a photo-ionised intergalactic medium (IGM) lies in the range $(1–5) \times 10^4$ K: the exact value depends on the spectrum of the UV background (which determines the mean energy per photo-electron when an ionisation occurs) and on the effects of adiabatic expansion, etc. An alternative idea, widely discussed in earlier years, is that the intergalactic medium might be hot enough to emit X-rays ($\gg 10^6$ K), and that collisional ionisation alone can maintain the neutral fraction below the requisite 10^{-6}. This idea now seems barely tenable. Whenever a hot thermal electron scatters soft photons – for instance those of the microwave background – the mean photon energy is raised by $\delta\nu/\nu = (kT/m_e c^2)$. The distortions of the microwave-background spectrum from an exact black body are now known to be below the level of 10^{-4}, and this severely constrains the thermal history of intergalactic gas with $T > 10^6$ K.

Inhomogeneously distributed gas: absorption lines

A uniform IGM has proved elusive, and its properties are now quite severely constrained. But even at the epoch probed by high-z quasars, we would expect the baryons to be concentrated in protogalaxies and other structures, and of these there is plenty of evidence.

Although quasar spectra show little evidence of any uniform opacity attributable to a smooth intergalactic medium, they reveal huge numbers of absorption features due to concentrations of gas at various redshifts along the line of sight.[7] Progress in interpreting these absorption features, depicted schematically in Figure 27, is fortunately not stymied by our poor understanding of the intrinsic properties of quasars: the quasar itself merely provides the light that serves as a probe of conditions in the intervening medium. In rare cases, along about one line of sight in ten, very broad and deep absorption lines are found, caused by neutral hydrogen (HI) with column densities as high as 10^{21} cm^{-2}. These probably implicate a protogalaxy or protogalactic disc. But weaker absorption systems are vastly more common. Even an HI column density as low as 10^{13} cm^{-2} produces detectable Lyman α, and

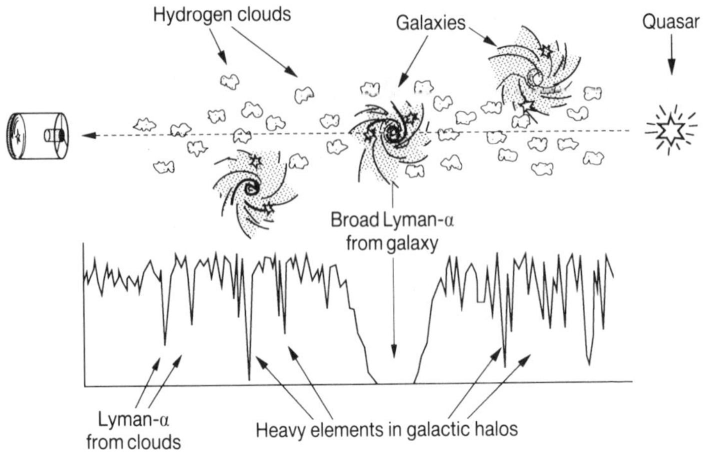

Figure 27 A schematic diagram indicating the wealth of information that can be revealed by absorption features in quasar spectra. Intervening galaxies can produce absorption lines due to numerous chemical elements; but there are also large numbers of smaller clouds which just produce hydrogen absorption in the Lyman lines. (Figure adapted from E. Stengler (unpublished).)

high-resolution spectra of high-z quasars reveal a forest of Lyman-α lines, implying that there are more than 100 'clouds' along every line of sight (see Figure 28). The 'Lyman forest' can be observed optically between redshifts of 1.8 and >4.5; smaller redshifts, for which the lines would be in the ultraviolet, have recently been studied with the Hubble Space Telescope. The 'forest' becomes thicker towards higher redshifts, implying that the relevant clouds are more common, larger, or less highly ionised at earlier epochs.

The relation of the inferred low-column-density clouds to galaxies is less clear. One attractive hypothesis is that they are due to gas falling into shallow potential wells associated with sub-galactic aggregates of dark matter (minihalos with virial

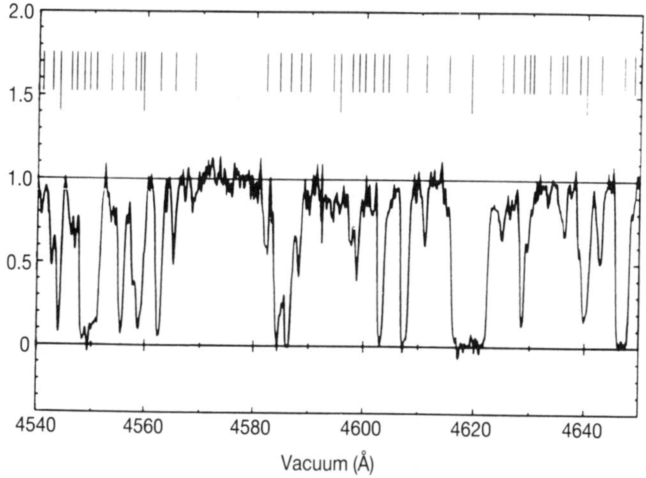

Figure 28 Segments of the spectrum of the quasar 0014+813 taken with high resolution, showing the number of lines in the 'Lyman forest' even within a range of 100 Å. (Data from Rauch, M. *et. al.* 1992, *Astrophys. J.* **390**, 387).

velocities V_c in the range 20–50 km s^{-1}) which should be much more common at high redshift than galactic-mass halos.[99] Any minihalo intercepting the light from a quasar would give an absorption line. The column density would be lower if the line of sight grazed the edge of a cloud than if it passed through the centre. The gas in these clouds, as in the general intergalactic medium, would be highly photo-ionised, so that the total column density would not be nearly as low as that in HI. An attractive feature of this hypothesis is that the very simplest assumption – that the confined gas has an isothermal distribution with total density proportional to r^{-2} (and therefore an HI density proportional to r^{-4}, since the recombination rate per unit volume goes as density squared) – predicts a number-versus-column-density relation going as $N^{-5/3}$, which agrees well with what is measured. But the assumptions of spherical symmetry and stationary virial equilibrium have limited validity. Outside the virialised region, the overdensity would drop gradually towards unity – moreover, the lines of column density below 10^{15} cm^{-2} would all come from these outer parts. A proper interpretation of the (much more numerous) systems with lower column density therefore requires us to consider the non-equilibrium dynamics of the outer regions, on scales of 10–50 kpc.[100]

The transverse scale of the clouds can be estimated from instances where the background quasar is gravitationally lensed by an intervening galaxy. The absorption spectra of both images, corresponding to lines of sight separated by around 10 kpc, are very similar.[101] This sets a lower limit to typical cloud sizes, which is compatible with the expected scale of minihalos.[100]

Proper modelling of the Lyman forest must await numerical simulations, incorporating gas dynamics and cooling, with enough resolution and dynamic range to handle these small-scale systems. The aim should be to account for the number-versus-column-density relation for the clouds, as well as any correlation between line width and column density. Higher-resolution spectra taken with the Keck 10 m telescope should soon offer an encouraging boost to these studies.

The z-dependence of the Lyman forest

Something else that must be explained – probably as a byproduct of a general theory for the emergence and build-up of structure – is the way the density of the Lyman absorption lines in redshift space depends on z, and therefore on cosmic epoch.

As hierarchical clustering proceeded, the number of potential wells into which gas falls (or in which gas is stably confined) would change. Minihalos form whenever low-mass systems turn around and virialise, but are eliminated when these merge to form larger halos with deeper potential wells. Numerical simulations do not yet have enough dynamic range to yield reliable numbers. However, analytic arguments[102] indicate that, in a standard CDM model, the minihalos should be thinning out, owing to mergers into larger systems with bigger V, at redshifts $z < 3$. The exact dependence is a function of the uncertain amplitude of the CDM spectrum, but it can be as steep as $(1 + z)$.

A second important effect is the change with redshift in the intensity and spectrum of the UV background J_{UV} from quasars and young galaxies. At redshifts $z < 2$, when the 'quasar era' is over (see Chapter 4), J_{UV} would decline. This would increase the neutral fraction in the gas at lower z, partly cancelling the first expected effect.

The epoch $z > 5$

When did reheating occur?

The lack of a 'Gunn–Peterson' cut-off in the spectra of even the highest-redshift quasars implies that the IGM must be almost fully ionised at least back to $z = 5$. There must therefore already have been at least one UV photon produced for each atom in the IGM. In cosmogonies such as 'standard CDM', the reheating of the IGM would be unlikely to have started much before the epoch $z = 10$. However, there are other models (cf. panel 3 of Figure 13) where 'first light' could have been much earlier; and conceivably there is some early heat input from (for instance) decaying particles.[103] Unless quasars are discovered at vastly greater redshifts than the current record, information will remain sadly lacking on the cosmic dark age, which stretched from the epoch of recombination to the stage when the first heat or ionising radiation was injected.

The epoch of reheating is important for interpretations of angular fluctuations in the microwave background.[12,63,104] I assumed implicitly on p. 56 that a photon has only a small probability of being scattered between the recombination era and the present: the microwave-background measurements then tell us about gravitational potential fluctuations and Doppler effects on a 'last-scattering surface' at $z = 1000$. However, there is at least a possibility that a dense IGM that was ionised early (at a redshift $z \gtrsim 50(\Omega_{IGM}/0.05)^{-2/3}$) would provide an opaque 'screen' at a lower redshift. This would affect interpretations of angular fluctuations in the microwave background; in particular, fluctuations on angular scales below 5 degrees (crucial for probing fluctuations on scales relevant to clusters and superclusters, and for discriminating among different cosmogonic models) would be smeared out.

Neutral hydrogen beyond $z = 5$

There is an interesting technique that could detect diffuse *neutral* hydrogen at $z > 5$, and thereby probe large-scale structure before the first galaxies or quasars 'switched on' and reheated the primordial material. This technique depends on studying the 21 cm line from diffuse atomic hydrogen. In terms of brightness temperature, this line would contribute much less than the 2.7 K microwave background. The line contribution is also much less than the background due to synchrotron emission from extragalactic radio sources. It may nevertheless be possible to pick out the 21 cm contribution, because of its characteristic angular structure, combined with fine structure in frequency space[105] (see Figure 29 and its caption).

The contribution to the radio-background temperature at $1420\ (1 + z)^{-1}$ MHz due to uniformly distributed hydrogen at redshift z is easily calculated to be

$$T_{HI} = 0.1(1 + z)^{1/2}\Omega_{HI} f \text{ K}. \tag{4}$$

The factor f depends on the 'spin temperature' T_s, which is defined by fitting a Boltzmann factor to the relative populations of the two hyperfine states. f is unity if the spin temperature is

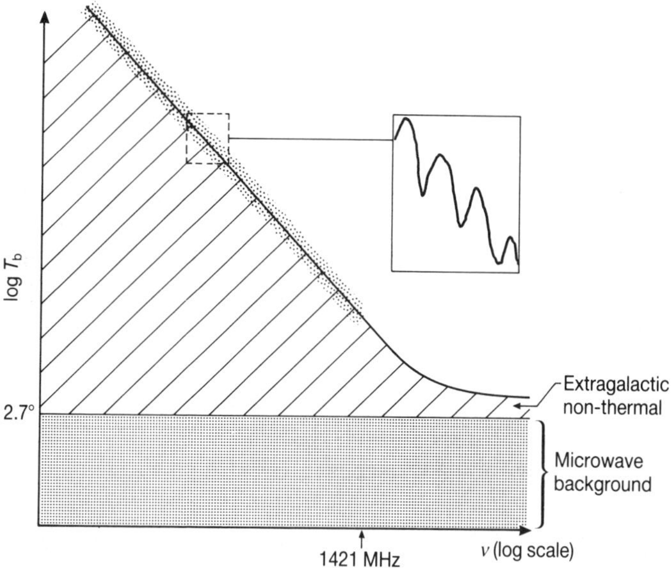

Figure 29 The dominant extragalactic backgrounds in the radio band are the primordial 2.7 K black-body radiation and the non-thermal synchrotron background, whose brightness temperature goes as $\sim\nu^{-2.7}$. At redshifts $z > 5$, the intergalactic gas may be mainly neutral (HI). If so, the high-z HI emits and/or absorbs via the 21 cm transition, and in consequence changes the background temperature. Although this effect would be undetectably small if the HI were smoothly distributed, any 'clumping' of the gas into incipient clusters would create spectral and angular structure in the background. By scanning in angle using a narrow bandwidth of frequencies, structures in the high-z neutral hydrogen could be detected. By comparing the angular structures seen in two 'maps' made at slightly different frequencies, one could distinguish between effects due to discrete non-thermal sources (for which the two maps would correlate) and those due to HI (where the maps would not correlate).

much higher than the radiation temperature. (Had there been no heat input at all into the primordial gas before the relevant epoch, the kinetic temperature, and therefore the spin temperature as well, could be *lower* than the radiation temperature, and the gas would show up in absorption against the black-body background radiation; f is then $-2.7(1+z)/T_s$). If some region along our line of sight had a higher density than average, or expanded at less than the mean Hubble rate, the contribution from the 21 cm line would be enhanced. For a linear fluctuation, the fractional enhancement is 5/3 $(\delta\rho/\rho)$, the extra 2/3 coming from the reduced expansion rate in an overdense growing perturbation, which further increases the hydrogen column density per unit redshift interval.

Large-scale inhomogeneities in HI at high redshifts would create spatial and spectral structure in the radio background. Though Equation (4) implies that these would be small compared to the total continuum radio background, they may be detectable by difference measurements, switching between nearby frequencies and directions. The expected fine structure in *frequency space* would distinguish the HI signal from that due to patchiness in the non-thermal synchrotron background (or, indeed, from angular fluctuations in the microwave background, though these are even smaller than would be detectable at these relatively low radio frequencies).

The most hopeful prospect for this kind of 'tomographic' exploration of protoclusters involves the Giant Metre-Wave Radio Telescope (GMRT) now being constructed by Swarup and his associates in India.[106] This instrument, when complete, will comprise an array of 34 dishes, each 45 metres in diameter. The dishes are not sufficiently well surfaced to be effective at high frequencies. However, the array will be eight times more sensitive than the VLA (Very Large Array) at 327 MHz. It will also operate in the 150–250 MHz band, where the artificial radio background, particularly at the GMRT's remote site, is especially low; this corresponds to redshifts between 6 and 8.5

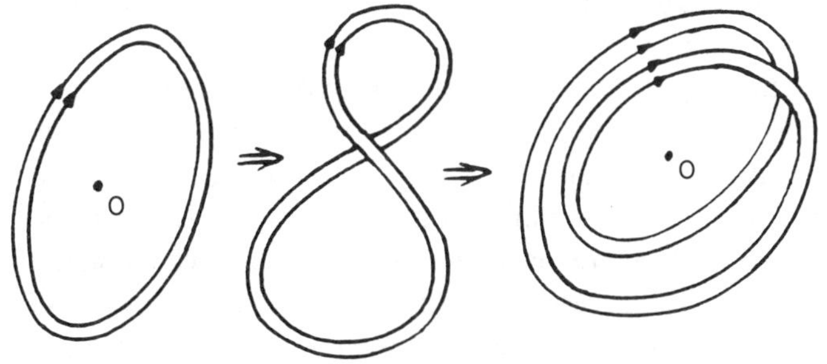

Figure 30 A single kinematic model (due to Zel'dovich) showing how magnetic flux can be systematically amplified by 'dynamo' effects, if magnetic loops are expanded and then twisted with a preferred helicity.

for the 21 cm line. The GMRT's planned sensitivity is such that protoclusters should be detected, if they have the properties predicted by some theories. Furthermore, the precursors of specially large concentrations such as the Great Attractor should reveal themselves in this way.

Magnetic fields

Primordial seed fields?

Cosmic magnetic fields pervade galaxies, and even clusters of galaxies,[107] and have widespread effects on gas dynamics and radiation processes. They probably owe their present strength to dynamo amplification, whereby kinetic energy is systematically converted into magnetic energy.[108] (Figure 30 shows how 'twisting' motions can enhance loops of magnetic flux.) But there must then have been an initial seed field – otherwise the dynamo process would have had nothing to feed on. It seems to be generally 'taken for granted' that the requisite seed field will be there. In many astrophysical contexts this confidence may be justifiable: if the dynamical (and amplification) timescale is short enough, there can be a huge number of e-foldings; a merely infinitesimal statistical fluctuation might then suffice. But the large-scale fields in disc galaxies seem to pose a less trivial problem. The amplification timescale, of the order of the orbital timescale, may be 2×10^8 years; even by the present epoch there has been time for only 50 e-foldings. A galactic field could not, therefore, have built up to its observed strength by the present day, unless the seed were of the order of 10^{-20}G – very weak, but not infinitesimal. Moreover, if it turned out that substantial fields existed even in high-z galaxies whose discs had only recently formed, the seed would need to have been correspondingly higher.

The question of how quickly cosmic fields gained strength is germane to several aspects of galaxy formation and evolution. *Star formation* would proceed differently (with regard both to its rate and the distribution of stellar masses) if there were no magnetic field: the field modifies the Jeans mass and helps collapsing protostars to shed angular momentum.[109] We cannot hope to model galactic evolution adequately without knowing when the field builds up to a dynamically important strength. (Moreover, even a weaker field may be significant through its influence on thermal conductivity etc.) If several galactic rotation periods elapsed before a dynamically significant field built up, then the oldest stars may well, for this reason alone, have a different distribution of masses. There is as much reason to believe that the absence of a magnetic field affects stellar masses as to believe that a lack of heavy elements (which affect cooling and opacity) would do so, though the quantitative nature of the effect is as uncertain in the one case as in the other.

Could a magnetic field have been created in the early stages of the big bang?[110] The ultra-early universe may have undergone a phase transition; and maybe this transition could (as in a cooling ferromagnetic material) spontaneously create a field. Because the relevant physics is exotic and poorly understood, we cannot rule this out. However, the natural correlation scale would be limited to the scale ct of the particle horizon, and this severely constrains a wide class of models. Suppose that, at a very early time t, some physical process generates an ordered field on a scale less than the horizon scale at t, whose strength is such that $B^2/8\eta = F(aT^4)$ (with $F < 1$), and that the universe subsequently expands according to the ordinary (decelerating) Friedmann equations. (This assumption can be made because the field, along with the background radiation, would be created after any inflationary phase was completed – any pre-inflationary magnetic field would have been exponentially diluted.) Then on a galactic scale we would expect an ordered field with energy density

$$\varepsilon_{\rm mag} = F(aT^4)\left(\frac{\text{mass within horizon when field is created}}{\text{mass of galaxy}}\right). \qquad (5)$$

A seed field of 10^{-20} G has energy density $10^{-29}(aT^4)$. But if the transition that created the field occurred at the GUT ('grand unified' theory) era (when the horizon was only large enough to encompass about 10^4 baryons), the ratio in the brackets in Equation (5) is of the order of 10^{-65}! So, even if the field had a high local energy density (and F was not very small), it would be primarily on such small scales that it would quickly decay, and there would be no chance of getting even 10^{-20} G on the scale of a protogalaxy.

This is a generic problem with attributing a cosmological origin to the field, even if a convincing microphysical mechanism could be found. (Of course, this problem would be surmounted if there were an overall cosmic anisotropy.)

(It is perhaps worth commenting parenthetically on how a primordial field, already present in the pre-recombination era, might affect the cosmogonic process. The constraints are summarised in Figure 30, in terms of the field's characteristic lengthscale. A field whose comoving strength was now $4 \times 10^{-10}\Omega_b G$, where Ω_b is the fraction of the critical cosmological density in baryons, would, at recombination (and at all later epochs until reheating occurred), have contributed more pressure than the baryons and electrons; it would therefore have affected the Jeans mass, and raised the minimum mass of the first generation of bound systems that would be expected in all 'hierarchical' models for the build-up of cosmic structure. Moreover, even a field too weak to affect the Jeans mass could still be cosmogonically important. This is because a field with characteristic scale l would (because of the inhomogeneous stresses) induce motions at about the Alfven speed on those same scales. Any resultant density fluctuations whose amplitude, at t_{rec}, exceeded 10^{-3} would have become nonlinear, via the ordinary gravitational instability, by the present epoch. Thus, even an intergalactic field as low as 10^{-13} G could have been cosmogonically significant if it dated from the pre-recombination era. The present constraints on an intergalactic field come from upper limits to intergalactic Faraday rotation. These depend on the field's correlation length, and are sketched in Figure 31).

Protogalactic batteries

If there is no significant primordial field, the 'seed' for dynamo amplification must have been created by a battery mechanism, which requires some large-scale vorticity. If the primordial fluctuations were irrotational (as they are in most models), then this would have had to await nonlinearities that lead to shock waves, or the formation of bound systems that exert tidal torques on each other. Compton drag can then gradually build up a current in a rotating protogalaxy. If plasma moves at a speed V relative to the frame in which the microwave background is isotropic, its motion would be damped out on a timescale $(m_p/m_e)t_{Comp}$, where $t_{Comp} = m_e c/aT^4$ is the usual Compton-cooling timescale for electrons. To couple electrons and ions, an E-field of strength $m_e V/et_{Comp}$ must maintain itself

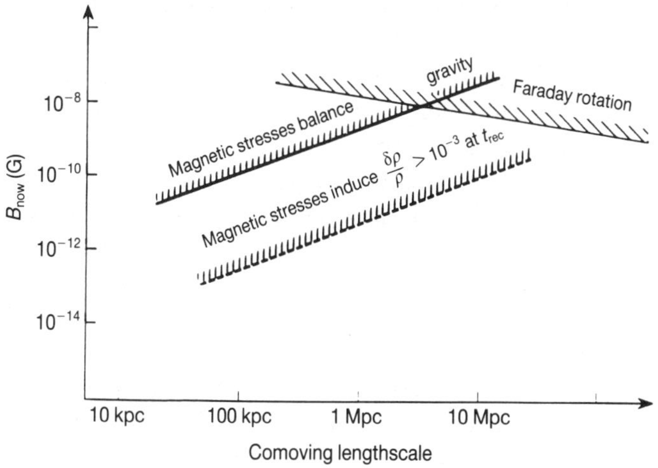

Figure 31 Constraints on the magnetic field on various length scales at the time of recombination t_{rec}. A field is cosmogonically important if it can generate density perturbations of amplitude $\sim 10^{-3}$ at t_{rec}, since these would have developed into gravitationally bound systems by the present time. See text for further explanation.

in the plasma. A protogalaxy of radius R rotating with speed V would be gradually braked by Compton drag, and the E-field within it (with, of course, non-zero curl) would build up a B-field at a rate $(m_e c^2 t/e t_{Comp})(V/R)$. For a protogalaxy at redshift $z = 5$, this process yields a field of the order of 10^{-21} G, which seems unpromisingly weak; we would indeed be impelled to explore other options if evidence emerged that even high-z galaxies had strong magnetic fields.

Magnetic fields from the first stars

Protostars condensing from the present-day interstellar medium start off with *too much* magnetic flux rather than too little

– the field has to diffuse out, probably by the process known as ambipolar diffusion.[111] But the field in a star at the *end* of its life may be insensitive to the conditions at its birth: even if a star initially had zero field, the battery process first discussed by Biermann[112] (which is more effective and rapid than the galactic-scale Compton-drag battery because of the much smaller scales involved in individual stars) could generate a seed field, on which dynamo amplification, by a huge number of factors of e if necessary, could operate. If such a star exploded as a supernova, then a wind spun off the remnant pulsar could pervade several cubic parsecs with a field of the order of 10^{-4} G (just as in the Crab nebula).[113] So the first few supernovae could have created a weak field throughout the galactic disc, even if a larger-scale battery had not already done so.

AGNs and radio lobes

The radio galaxy with the highest redshift so far discovered,[114] 4C 41.17, has radio lobes 30 kpc in size, containing ordered fields of 10^{-5}G, implying a flux of the order of 10^{41} G cm^2. Its redshift, $z = 3.8$, corresponds to a cosmic epoch when the universe (if it is described by the Einstein–de Sitter model) was only about a tenth of its present age.

Radio galaxies like 4C 41.17 may well have formed exceptionally early, when the formation of typical galaxies (especially those with discs) still lay in the future. The fields in the lobes of radio galaxies could have been generated in the active nucleus of the associated galaxy and expelled along collimated jets (resembling a scaled-up and directional version of the relativistic pulsar wind that generates the Crab nebula's field). Thus, a radio galaxy's field, like that in a supernova remnant, can be accounted for even if the progenitor central object had zero field when it formed. And the radio lobes could, in their turn, 'seed' galactic discs, if the lobe material were subsequently mixed into a larger volume.

What is the most likely origin of an adequate seed field?

The seed field for the *galactic* dynamo[115] poses a more challenging question than the seeding of smaller-scale cosmic dynamos because the galactic timescale is so long, and the amplification correspondingly slow. (And I have assumed, of course, that the galactic dynamo mechanism is indeed efficient – the problem is obviously far worse if it is not.) There are as yet no firm grounds for expecting significant fields in the ultra-early universe – indeed there are good reasons for expecting the large-scale components of any such field to be uninterestingly small. And the galactic-scale batteries where Compton drag provides the e.m.f. would be barely enough to yield an adequate seed. More promising, in my view, are the two options depicted in Table 2, either of which could yield 10^{-9} G (and which are not mutually exclusive).

The build-up of a galactic magnetic field depends on how strong the seed field is and when it was generated. Because of its importance in star formation, we have little chance of really understanding what a high-redshift galaxy should look like until these issues have been given a good deal more attention by experts in cosmic magnetism.

Cosmic strings

We cannot tell whether magnetic fields were 'seeded' in the ultra-early universe – if they were, their presence would tell us something about exotic physics. What would be really exciting would be some relic that unambiguously revealed something about such physics. Cosmic strings would be a spectacular example.

Cosmic strings are one-dimensional topological defects that could have formed in the early universe as relics of vacuum phase transitions.[116] They arise in many, though not all, gauge

Table 2

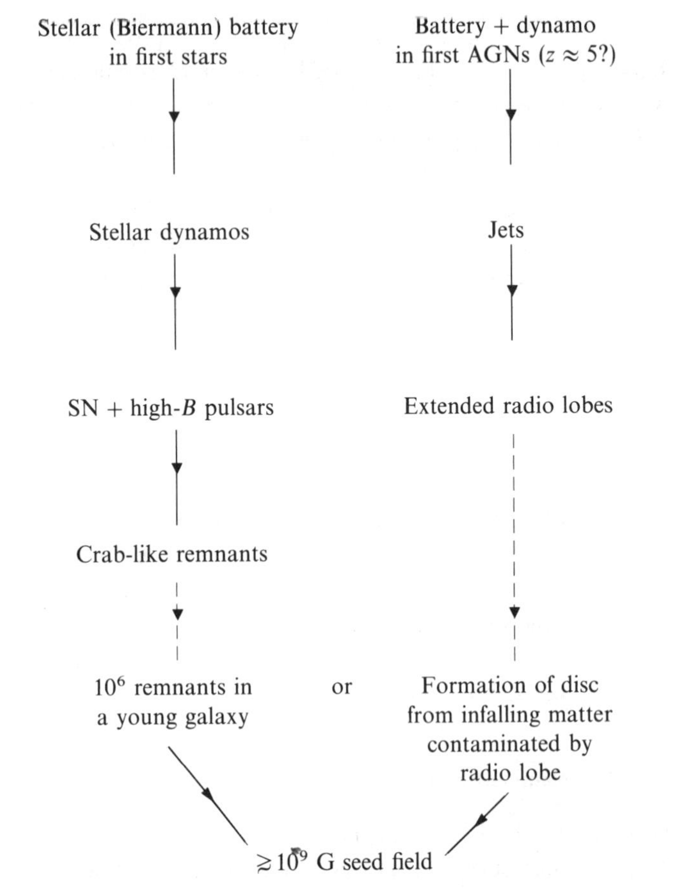

theories in which the original symmetry group is spontaneously broken. They are lines of trapped energy with remarkable dimensions: twenty powers of ten thinner than the size of an atom, but with a mass of 10^{17} tonnes per metre. The mass is characterised by a dimensionless number $G\mu$, a measure of the number of Planck masses $[(hc/G)^{1/2}]$ per Planck length $[(Gh/c^3)^{1/2}]$. Other topological defects – domain walls and monopoles – would be a cosmological embarrassment. Strings, on the other hand, would be welcome to many cosmologists.[117]

During an early phase transition, a network of strings might form, whose subsequent evolution can be modelled. When a long string intersects itself, a closed loop breaks off. At late times, we expect a few 'open' strings stretching right across the observable universe, together with an array of closed loops extending down to small scales. The gravitational effects of stings have been proposed as 'seeds' for galaxy and cluster formation.[117] This would require $G\mu$ to be a few times 10^{-6}. Sting-induced structure formation would have two distinctive features: it would be induced by fluctuations that were highly non-Gaussian, and there would be a natural predominance of large-scale linear features. Strings are in principle detectable through the distinctive gravitational lensing patterns that they could produce, or through the characteristic sharp-edged imprints they would create in the microwave background.[118] But their most conspicuous effects would be the gravitational waves they would generate.[119]

Loops of cosmic string, flailing around at speeds $\sim c$, emit gravitational waves; the resultant energy loss causes them to shrink, and eventually disappear. Neither gravitational wave detectors on the ground, not the space interferometers that could detect massive coalescing black holes (see p. 83) would be sensitive enough to reveal these gravitational waves. However, the radiation from decaying string loops would have a range of periods extending up to years, and this ultra-low-frequency background would manifest itself via observations of a kind that might seem, prima facie, entirely unrelated – precise recording, by radio astronomers, of the periods of pulsars in our own galaxy.

One particular class of pulsars, the so-called millisecond pulsars spinning at several hundred revolutions per second,

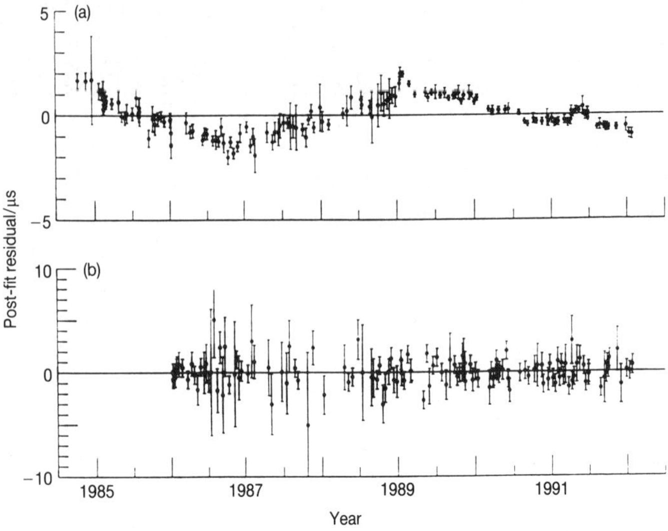

Figure 32 Timing residuals for two 'millisecond' pulsars: (*a*) 1937 + 21 and (*b*) 1855 + 09, from work by Taylor and his collaborators. Note that these objects are steady clocks, with a precision of a fraction of a microsecond over several years. These pulsars are sensitive detectors of gravitational waves with periods 1–10 years; the steadiness of these periods therefore sets a constraint on cosmic strings. (From Taylor, J. H. 1992, *Phil. Trans. Roy. Soc.* **A341**, 117.)

have such sharp and steady pulses that the timing residuals can be as small as a fraction of a *micro*second.[120] If the space between us and a pulsar were perturbed by gravitational waves, 'timing noise' would be introduced. The pulsar provides the greatest sensitivity as a probe for waves whose period is comparable with the length of time for which observations have been made – this is of the order of 10 years. The pulsar clocks (see Figure 32) are 'steady' to a few parts in 10^{15}. The inferred limit on the gravitational-wave background precludes a strong network with $G\mu$ more than a few times 10^{-7}. The main imprecision in the limit stems from theoretical uncertainty about how a string network would evolve,[117] and therefore in the space density of loops of the appropriate size to produce ten-year-period gravitational waves.

In view of the long-standing failure of astronomers to determine Hubble's constant to anything much better than 50 per cent accuracy, it is gratifying to record an astronomical phenomenon that can be measured to 15 significant figures. It is indeed astonishing that the motions of a tiny neutron star thousands of light-years away can be measured so precisely that, within a timebase of a few years, radial velocity variations of a few microns per second – the speed of the hour-hand on a watch – could be discerned.

Strings could of course still exist, despite these gravitational-wave limits, if $G\mu$ were smaller. They would then be of less direct interest for galaxy formation, but still important as fossils of an early phase transition. The key role of these phase transitions (see Chapter 6) is as a 'driver' for inflation. They remind us that Gaussian adiabatic perturbations are not the only conceivable initiators of structure formation.

6

Some fundamental questions

Gravity

A word here about gravity, which holds together galaxies, as well as individual stars. The weakness of gravity on the *micro*physical scale is reflected by the fact that the 'gravitational fine-structure constant', $\alpha_G = Gm_{proton}{}^2/\hbar c$, is only 10^{-38}. Because α_G is so small, gravitational binding energy only becomes significant when huge numbers of nuclei are packed together. When N atoms are packed together within a radius proportional to $N^{1/3}$ (i.e. at fixed density), the gravitational binding energy per atom goes as $N^{2/3}$; gravity starts off with a 'handicap' of 10^{38}, so it only 'catches up' when N is as large as $10^{38 \times 3/2} = 10^{57}$. Figure 33 (and caption) illustrates this, and shows clearly that there are basic physical reasons why aggregates of around $\alpha_G{}^{-3/2} = 10^{57}$ baryons should behave as stars ('gravitationally bound fusion reactors'). A similar style of argument shows that stellar lifetimes of stars exceed microphysical timescales by the first power of α_G: these lifetimes can be expressed in terms of t_{Sal} (see Equation (3)), which explicitly involves α_G. The universe had to be old by the time we came on the scene; it therefore (if it is homogeneous) has to be large as well. Another corollary of the largeness of $\alpha_G{}^{-1}$ that emerges clearly from Figure 33 is that the Planck scales, where gravity and quantum effects overlap, lie many powers of ten away from the 'ordinary' scales confronted in the laboratory, or even in astronomy. If a Friedmann cosmological model could be validly extrapolated back to the Planck time $(G\hbar/c^5)^{1/2} = 5 \times 10^{-44}$ s, the first millisecond would contain 40 'decades' of logarithmic time.

Structures large enough to be controlled by gravity have a peculiar property: when they lose energy, they heat up; their specific heat is negative. For instance, if its radiative losses were not compensated by nuclear fusion, the Sun would contract and deflate, but would end up with a hotter centre than before: to establish a new and more compact equilibrium where pressure can balance a (now stronger) gravitational force, the central temperature must rise. Gravity drives things further from homogeneity – further from a global state of thermal equilibrium. It is because of gravity that regions only slightly overdense in the early universe would have suffered extra deceleration, lagging behind more and more until their expansion eventually stopped. This gravitational instability transformed small fluctuations or ripples in the early universe into galaxies, clusters, and superclusters (see Chapter 3).

If one wished to summarise in just one sentence what has been happening since the big bang, the best answer might be to take a deep breath and say: 'Ever since the beginning, the 'antithermodynamic' effects of gravity have been amplifying inhomogeneities, and creating progressively steeper temperature gradients – a prerequisite for the emergence of the complexity that lies around us 10 billion years later, and of which we are a part.'

The *ultra*-early universe

The progress in understanding how galaxies and clusters formed brings new questions into focus. Why was the universe set up with the observed mix of particles and radiation? Why does it have its distinctive geometrical structure – the overall homogeneity that makes cosmology tractable combined with the small-amplitude metric fluctuations or ripples without which our universe would still be featureless? The answer to these questions lies in the *very* early universe.

I emphasised in Chapter 1 that there were quite firm grounds for extrapolating back to 1 second, when the temperature

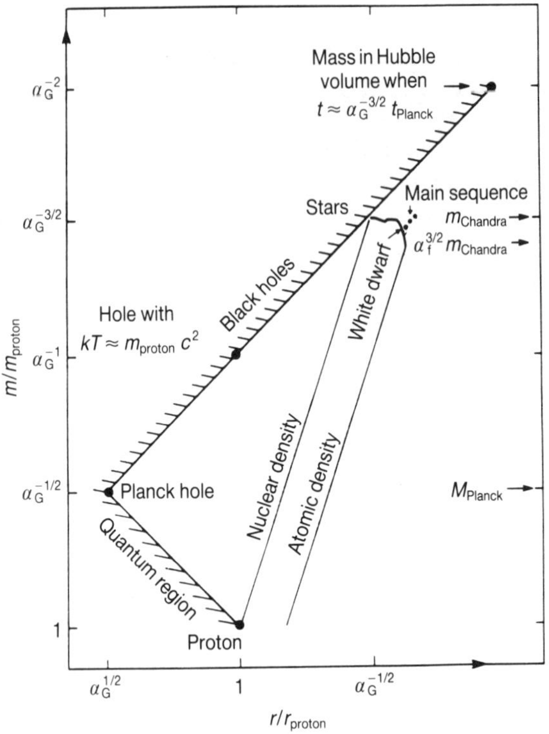

Figure 33 This diagram summarises the physics of stars, planets, etc. in a mass–radius plot. The most striking feature is that significant phenomena occur for masses related to m_{proton} by simple powers of α_G. The Planck mass, for which Compton and Schwarzschild radii are equal, is $\alpha_G^{-1/2} m_{proton}$; the corresponding radius is the 'Planck length' $(G\hbar/c^3)^{1/2}$. It is because these dimensions lie so far from either laboratory or astrophysical relevance that our scientific progress is not primarily impeded by ignorance of quantum gravity. A mass $\alpha_G^{-1} m_{proton}$ corresponds to a black hole whose radius is the size of a proton: such a hole has a Hawking temperature $kT \approx m_{proton} c^2$, and radiates in a time of the order of $\alpha_G^{-3/2} t_{Planck}$ (i.e. a stellar lifetime, t_*). Stellar masses are $\sim \alpha_G^{-3/2} m_{proton}$. The mass scale $\alpha_G^{-2} m_{proton}$ is also of significance as being the mass within a Hubble volume for a flat

would have been 1 MeV. At that time, when neutrons and protons started synthesising the light elements, the densities and energies were modest and the microphysics was 'standard'. But at still earlier times conditions would have been more extreme. The universe would have exceeded nuclear density for the first 10^{-3} seconds. When $t = 10^{-12}$ s, each particle had energies exceeding 1 TeV – beyond the range of present terrestrial accelerators. Twenty four further decades of logarithmic time take us back to the conditions when the temperature was 10^{15} GeV (relevant to unified theories) and 8 more to the era of quantum gravity when the Hubble radius was equal to the Planck length ($(G\hbar/c^3)^{1/2} = 10^{-33}$ cm). Note that the Planck time is less than the age of the universe when $kT = m_{\text{proton}}c^2 = 1$ GeV by a factor which is huge because it is essentially equal to the number $\alpha_G{}^{-1}$ mentioned above.

One of the 'initial conditions' for structure formation is the baryon-to-photon ratio η. This is a small number of the order of 10^{-9}. When $kT > m_{\text{proton}}c^2$, the asymmetry between protons and antiprotons (or quarks and antiquarks) would have been small, of the order of η. As the universe cooled, the pairs would have annihilated into radiation, leaving only a small 'unpaired' fraction of the order of η. What is the explanation for this crucial number? If it were zero, the universe would, after 10 billion years, be almost pure radiation. The smallness of the number compared to unity has led most theorists to interpret it as a 'baryon favouritism' in the ultra-early universe. The smallness of the asymmetry $(n_b - \bar{n}_b)/(n_b + \bar{n}_b) = \eta$ is, in many theories,

---

Caption for Figure 33 (*cont.*)

Friedmann universe whose age is $\sim t_*$. (Galactic masses could also be included in such a diagram, but as discussed in Chapters 2 and 3 their physics is less clearcut.) This diagram emphasises that it is only because α_G^{-1} is so huge that so many powers of ten separate the microphysical from the astrophysical scales.

related to the small degree of CP (charge/parity) violation in weak interactions. I will not address these ideas further, except to mention two points:

(i) The asymmetry was determined at a very early time when the mass within a 'Hubble volume' (the so-called 'horizon scale') was small. Though η might display *very*-small-scale spatial variations (associated, for instance, with topological defects), it should be uniform on all astrophysically interesting scales. Galaxy formation from so-called primordial isocurvature baryon fluctuations in η (which would be frozen-in at constant amplitude until recombination) has been discussed,[121] but there is no physically well-grounded model for generating isocurvature fluctuations on sufficiently large scales. This is the rationale for favouring the idea that structures originate from curvature perturbations which do not affect the photon–baryon ratio.

(ii) An analogous 'favouritism' for matter over antimatter might also occur for non-baryonic dark matter in the form of WIMPs. The number which survive may therefore not be set by the fraction that escape annihilation, but by some mechanism similar to that responsible for the cosmic baryon excess.

Ratios like η are universal cosmic numbers, in the sense that they may take the same value throughout our observable universe; they are determined essentially by the outcome of as-yet-unquantifiable local microphysical processes (collisions, annihilations, etc.), which take place while the primordial material expands and cools through some critical temperature range. But this in a sense begs a still more fundamental question: why does the universe have the same expansion rate everywhere?

Flatness and the horizon problem

One basic mystery is why the universe is still, after 10^{10} years, expanding with a value of Ω not too different from unity. We do not know the long-range forecast, and the contrasting scenarios of perpetual expansion versus recollapse to a big crunch seem very different. But if the expansion was 'set up' at some early time, it looks surprising that the universe has neither collapsed long ago, nor is expanding so fast that its kinetic energy has overwhelmed the effect of gravity by many powers of ten. It seems that the conditions would need rather careful tuning to get to where we are now.[122] This is called the *flatness problem*.

The related *horizon problem* is even more perplexing. Why should the universe have such simple dynamics that it can be described by a single scale factor $R(t)$? More degrees of freedom would seem to be open to a universe that was wildly inhomogeneous and anisotropic. So why were all parts synchronised to start expanding in the same way and obeying the same dynamics? Somehow all parts of the universe that we can observe seem to have homogenised themselves. But in the standard models causal contact was worse in the past. At early times, the scale factor R goes as t^n, where $n = 2/3$ when pressure is negligible compared with ρ/c^2, and $n = 1/2$ in the early radiation-dominated phases. When the universe was compressed by a factor x (i.e. when R was x times smaller than R_{now}), its expansion timescale was less than it now is by $x^{1/n}$. The horizon problem arises because $n < 1$. At an early time, when R was smaller, everything was closer together. One might have thought that this would improve causal contact (by, for instance, permitting easier exchange of light signals, pressure waves, etc.). But the causal contact was actually *worse*, because the time available was shorter by an even bigger factor.

The horizon problem can be illustrated by a simple example. Consider a galaxy 10^9 light-years from us. Since the age of the universe is $(1\text{–}2) \times 10^{10}$ years, there would be time for 10–20 signals to be exchanged. At the epoch of (re)combination, when R was $10^{-3}R_{now}$, the galaxy would have been only 10^6 light-years away – a thousand times closer. But since R would go roughly as $t^{2/3}$ during the relevant period, the universe at (re)combination was then expanding on a timescale that was faster not merely by a factor 10^3 but by $10^{4.5}$. There would therefore, by then, not have been time to exchange *even one* signal with the other galaxy.

Two galaxies that are destined to be 10^9 light-years apart at the present epoch would, if they lay on the 'last-scattering surface' at $z = 1000$, be separated on the sky by about $5\ \Omega^{1/2}$ degrees. So why, when we observe the microwave background from two directions separated by large angles, do we find that the temperature is almost the same?

Inflationary models

The horizon problem arises because gravity *decelerates* the cosmic expansion – i.e. $R \propto t^n$, with $n < 1$. The proposed solution is to postulate an *accelerating* phase of exponential expansion which *preceded* the $R \propto t^n$ phase. In an accelerating universe, causal contact would have been better at earlier times, so remotely separated parts of our present universe could have synchronised and coordinated themselves early in the inflationary phase. The details of 'inflation' are of course speculative, because they are sensitive to physics at ultra-high energies which is almost completely unknown. But the generic idea of inflation[123] is compellingly attractive because it suggests solutions to the flatness and horizon problems. Indeed in a sense it suggests *why* the universe is expanding – something which seems otherwise just a part of the 'initial conditions'.

According to inflationary models,[123–125] our universe is as big as it is, and is expanding the way it is, because there was an exponentially expanding era, probably before the time when $kT = 10^{15}$ GeV (when the expansion timescale was 10^{-36} s). If we extrapolate the Friedmann equations back to that time, we find that our currently observable universe would then have been a few centimetres across (see Figure 34 and caption). Though this seems small, it is exceedingly large compared with the scale ct across which causal contact could be maintained:

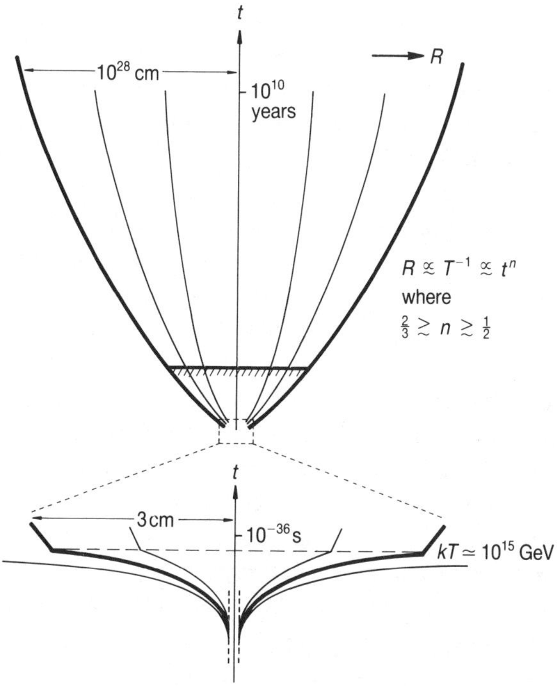

Figure 34 Schematic illustration (with time plotted vertically, and radial scale horizontally) of how the cosmic expansion is decelerating. Causal contact was worse in the past because of this deceleration – for instance, at the epoch of recombination, objects were separated by 1000 times less than at present, but the expansion timescale (and therefore the time available for exchanging signals) was then smaller not merely by 1000 but by about 30 000. When kT was 10^{15} GeV, everything within our *present* Hubble radius would have been squeezed down to a scale of a few centimetres. However, the effective Hubble radius *then* was only about 10^{-26} cm. To account for the present homogeneity of the universe, inflationary models postulate that there was, at a time still earlier than this, a period when R expanded at an exponential rate, by a factor much larger than 10^{26} (i.e. $\gg 60$ powers of e).

this would only have been 3×10^{-26} cm. But our observable universe could, nevertheless, have emerged from a causally connected domain no bigger than the Planck scale if a preceding era of exponential expansion had been maintained for more than about 60 'e-foldings'.

The fluctuations from which clusters and superclusters form, and the even larger ones spread across the sky whose effect on the microwave background has been measured, are, if these ideas are right, a genuine quantum phenomenon from an ultra-ancient epoch when the Hubble radius was only a few powers of ten bigger than the Planck length. The Harrison–Zel'dovich spectrum arises naturally. More precisely, one expects a slightly tilted spectrum with a slow logarithmic rise in amplitude towards larger scales. This is because growth does not completely 'shut off' when the scales inflate beyond the event horizon, and the larger scales have expanded by more powers of e from the Planck scale.[125,126]

The 'magic number' $\psi = 10^{-5}$ characterising the 'roughness' of the universe is not clearly understood, and depends on physics that is still speculative. But specific models of inflation make distinctive quantitative predictions about the 'tilt' of the fluctuations, and also about such things as the ratio of scalar and tensor modes; these are both being probed observationally by studies of large-scale clustering, and of fluctuations in the microwave-background temperature on various angular scales.[127] We can therefore look forward soon to empirical probes of the quantum-gravity era: even if we do not know the appropriate physics, we can calculate the quantitative consequences of a specific model, compare these with the observations, and thereby at least constrain the possible physics.

One generic prediction of inflation is that Ω should be almost exactly unity. In most versions of the theory, the exponential growth, once started, is likely to 'overshoot', stretching any small part of an initial chaotic hypersurface so large that it becomes essentially flat over any scale we can now observe. For inflation to yield (say) $\Omega = 0.2$, the inflation factor would have to be 'just' 10^{29}, making the Robertson–Walker curvature scale comparable with the present Hubble radius. This would demand some coincidence. But there would then be an additional requirement that appears still more contrived: our present universe would have to arise from a segment of the initial hypersurface with the seemingly very special property that its curvature was uniform to a few parts in 10^5; otherwise, curvature fluctuations surviving from the pre-inflationary period would create much more prominent dipole and quadruple (and other large-angular-scale) fluctuations in the microwave background than are actually found. Our universe would have had to inflate from a segment of the initial hypersurface that was special, rather as a sphere would seem special if its surface irregularities amounted to no more than 10^{-5} of the mean curvature.

This is the basis for the strong theoretical 'prejudice' for $\Omega = 1$. (More precisely, since the Harrison–Zel'dovich spectrum should extend to scales exceeding the present Hubble radius, we would expect $|\Omega - 1|$ to be of the order of 10^{-5}.) As discussed in Chapter 3, the evidence on this issue is still unclear: there is enough dark matter to contribute 0.1–0.2, but the best empirical support for $\Omega = 1$ comes from the 'large-scale streaming motions' of galaxies relative to the Hubble flow, which (at least provisionally) indicate inhomogeneities in the gravitational field on the scale of superclusters large enough to require still more dark matter.

Inflationary models are now 15 years old. The general idea has survived, but its detailed implementation has gone through several fashions. There is no consensus on the link between any specific unified theory and the inflationary phase transition (earlier ideas having fallen from favour). However, it is no coincidence that the inflationary-universe idea first came to prominence at the time when grand unified theories were being seriously discussed. Baryon non-conservation is essential in any inflationary model. It is no good inflating the universe to an enormous volume if it cannot afterwards be populated with baryons; this requires that the phase transition at the end of inflation should reheat the universe enough to ensure that a baryonic favouritism $(n_b - \bar{n}_b)/(n_b + \bar{n}_b) = \eta$ is established throughout the inflated volume.

Many theorists now favour the idea of *chaotic inflation*.[128] In this model, there is not necessarily, as depicted in Figure 35, an initial simple singularity, with the inflationary phase being an intermediate interlude. Linde envisages a more complex situation where the collapse of some region triggers creation of a new cosmos. If our universe is almost flat, it extends vastly beyond our present horizon of 10–20 billion light-years, and is destined to expand much more, with many more galaxies coming into view. It may still eventually recollapse, but only after expanding by a further factor 10^{10^5}! But what we call '*our*' universe may be just one domain, or one phase, of an eternally reproducing cycle of different 'universes'. These are not now in causal contact, but can be traced back to common ancestors. Moreover, the phase transitions, compactification, etc. may have proceeded differently in all of the 'universes', only some ending up (like our own) as propitious locations for complex astrophysical evolution.

This scenario, setting our entire observable universe in a magnificent infinite ensemble, is certainly on the 'way-out' fringe of speculation. Indeed, most current specific models for the ultra-early universe have, it is probably fair to say, a short shelf-life. The physics of the first 10^{-35} seconds, perhaps even the entire first microsecond, is as uncertain today as was the physics at $t = 1$ s when Gamow and other pioneers first explored the cosmological origin of the elements. Their ideas were put on a firm footing within a decade or two. Maybe we can share similar hopes about a symbiosis between ultra-high-energy physics and cosmology in the next decade.

Primordial nucleosynthesis, occurring when $t = 1$–100 s and

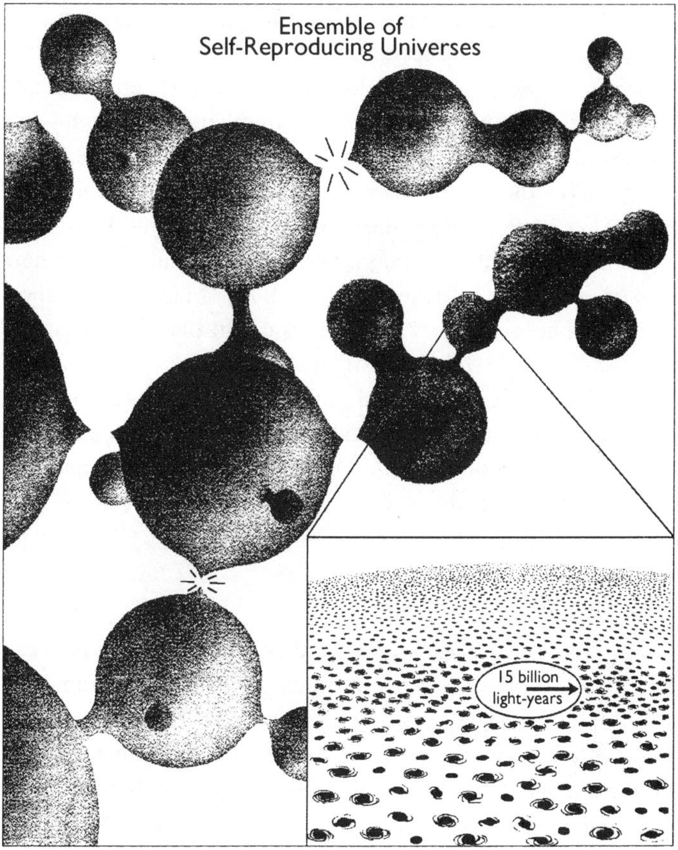

Figure 35 A very schematic sketch of Linde's concept of 'external inflation'. Our own 'universe', which itself extends vastly beyond the present horizon of $\sim 10^{10}$ light-years (inset) is just one part of an infinite ensemble. (From Mallove, E. F. *Sky and Telescope* Sept. 1988, p. 256. Diagram © 1988 Sky Publishing Corp. Reproduced with permission.)

$kT = 0.1$–1 MeV, involved physical processes (low-energy nuclear physics, etc.) that could be explored experimentally. In contrast, the energies and densities at ultra-early eras relevant to determining fundamental cosmic numbers like η and ψ are too extreme to be simulated terrestrially, even in accelerators. That makes the new challenge more daunting. On the other hand, that same circumstance provides an extra motivation, in that the early universe may offer the only real tests of new unified theories because it is the only place where their distinctive consequences are manifested. In the 1950s, cosmology was outside the mainstream of physics – only a few 'eccentrics' like Gamow paid any attention to it. In contrast, cosmological issues now engage the interests of many leading 'mainstream' theoretical physicists. And that surely gives us grounds for optimism.

Concluding homily

Cosmology is a subject which straddles the boundary between fundamental physics and what might be called environmental science. It consequently confronts us with two contrasting styles of problem. Some aspects of cosmology resemble particle physics, a subject whose practitioners aspire to theories with only a few parameters, which can be accurately compared with the data.

There are essentially just four basic 'cosmic numbers':

(i) α_G which measures the weakness of gravity compared to other forces; the fact that this is so enormously different from unity permits a huge hierarchy of particle masses, and of cosmic scales;

(ii) the overall curvature (related to the mean cosmic density parameter Ω), which inflationary models predict to be zero;

(iii) the baryonic density, described by the parameter η; and

(iv) the curvature fluctuation amplitude ψ, which in inflationary models is essentially just a single number (and which is observationally constrained to be close to 10^{-5}).

There is real hope of deducing these four cosmic numbers from fundamental physics. Moreover, when supplemented by knowledge of what the dark matter is, which may itself soon come from basic physics, these numbers suffice to determine the main features of our present universe – the epoch of galaxy formation, the large-scale structure, the light elements, etc.

But the genesis of an individual galaxy involves dissipative gas dynamics, star formation, and the feedback from stars and supernovae. These are all complicated and messy processes. Models of galaxy formation will never be as 'clean' as the theories that particle physicists aspire to – there will always be an element of 'parameter fitting' guided by the observations. The problems are so intermeshed that we will not really solve any until the whole picture comes into sharper focus. For instance, we cannot test theories of galaxy formation and evolution until we understand the gas dynamics of star formation, and the possible role of active nuclei, as well as the exotic physics of the initial fluctuations.

As observations and modelling advance in tandem, we can realistically hope for new insights that will reduce our current bafflement, and narrow down the range of hypotheses regarding dark matter, structure formation, and so forth. These insights will not resemble a new theory in fundamental physics: they will be more like a good theory in geophysics – plate tectonics for instance. This is a unifying idea that gives insight into previously unrelated facts, but it is no disparagement of plate tectonics that it cannot predict the shape of the continents.

When the COBE results on anisotropies in background radiation were announced, the English newspaper *The Independent*, under the heading 'How the universe began', filled most of its front page with an elaborate 'time chart' depicting the history of the universe from 10^{-44} seconds to the present. Anyone who has dealt with the media knows two things about them. First, their interest is somewhat capricious and arbitrary. Second, if interest gets above some threshold level, if feeds on itself and amplifies. That certainly happened here. The discovery that triggered this, one of the two major results from the COBE satellite, was undoubtedly a technical triumph, but the 'hype' was disproportionate.

Although anything that stimulates public interest in science is welcome, I am somewhat uneasy about how cosmology is sometimes popularised. First, if we claim too often to be stripping the last veil from the face of God, or making discoveries that overthrow all previous ideas, we will surely erode our credibility. It would be prudent, as well as seemly, to rein in the hyperbole a bit. Second, one should not conflate things that are quite well established with those that are not yet in that state.

This distinction is brought out by Figure 36, which divides the history of a Friedmann 'hot-big-bang' universe into three parts. Part 1 is the first millisecond. Many key features of the present universe – the basic mix of particles and radiation, and the fluctuations from which cosmic structures emerged – are legacies of this phase, which sets the initial conditions for phase 2.

This second phase runs from 10^{-3} seconds to the epoch of recombination at about 10^{13} seconds. It is an era where cautious empiricists like myself feel more at home. The prevailing temperatures are below 100 MeV, and the densities (even when radiation is included) are far below nuclear densities, so the microphysics is no longer uncertain. Quantitative predictions based on well-known physics are possible: for instance, the standard fireball model, for an appropriate choice of baryon density, yields gratifying agreement with the observed abundances of the light elements. These elements, and the background radiation itself, offer our empirical basis for this second phase.

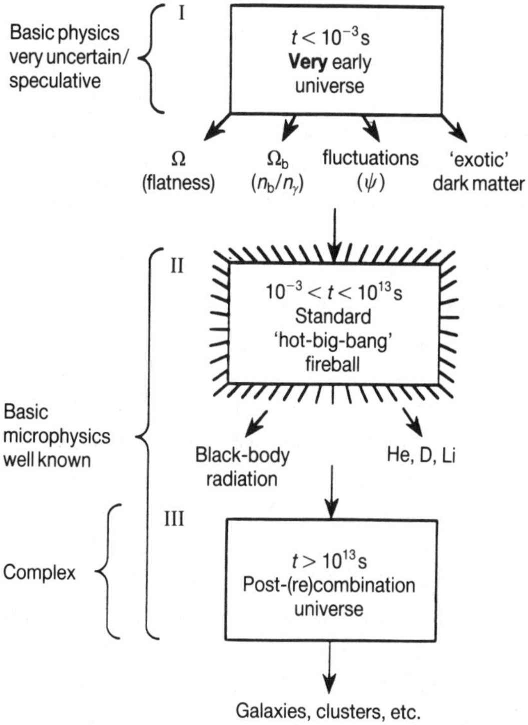

Figure 36 The history of a standard 'hot-big-bang' universe divided into three parts. See text for further explanation.

At some stage after the era of recombination, the present cosmic structures condensed from the amorphous fireball. The basic *micro*physics remains uncontroversial. But the simplicity ends once nonlinearities develop and bound systems form. Gravity, gas dynamics, and all the physics in every volume of Landau and Lifshitz, combine to initiate the complexities we see around us and are part of.

We have firm empirical support (and a firm link with 'known' physics) when we talk about $t \gtrsim 10^{-3}$ s and $kT \lesssim 100$ MeV –

i.e. primordial nucleosynthesis and what comes after. But we are on much shakier ground when our extrapolations venture right back into the first millisecond, and we should not disguise this. When popular presentations blur the difference between phase 1 and phase 2, there is a risk that readers will either accept speculations about the ultra-early universe overcredulously, or (if they are more sceptical) will fail to appreciate that some parts of cosmology (those pertaining to phases 2 and 3) have a firm and quantitative empirical basis.

But despite this 'health warning' my conclusion is an upbeat one. It is remarkable that telescopes can now survey 90 per cent of cosmic history, that we are starting to model in detail how the present cosmic structures evolve from amorphous beginnings, and that other techniques can probe still earlier epochs, and the nature of the dark matter. This progress brings into sharper focus a new set of fundamental problems about the very early universe whose solution demands new ultra-high-energy particle physics and unified theories. We can look forward to a challenging intellectual symbiosis between these subjects. Previously speculative parts of cosmology are now coming within the framework of serious physics. I hope these lectures have at least given some flavour of the progress and the prospects.

References

1. Friedmann, A. 1922, *Z. Phys.* **10**, 377.
2. Hubble, E. 1929, *Proc. Nat. Acad. Sci.* **15**, 168.
3. Bondi, H. & Gold, T. 1948, *Mon. Nat. Roy. Astron. Soc.* **108**, 252. Hoyle, F. 1948, *Mon. Nat. Roy. Astron. Soc.* **108**, 372.
4. Sandage, A. 1961, *Astrophys. J.* **133**, 355.
5. Ryle, M. 1958, *Prov. Roy. Soc. A* **248**, 289.
6. Schneider, D. P., Schmidt, M. & Gunn, J. E. 1991, *Astron. J.* **102**, 837.
7. For a review see Blades, C., Turnshek, D. & Norman, C. A. (eds.) 1988, *Quasar Absorption Lines* (Cambridge University Press).
8. Penzias, A. & Wilson, R. W. 1965, *Astrophys. J.* **142**, 419.
9. Mather, J. C. *et al.* 1990, *Astrophys. J.* (*Lett.*) **354**, L37. Mather, J. C. *et al.* 1994, *Astrophys. J.* **420**, 439.
10. Smoot, G. F. *et al.* 1992, *Astrophys. J.* (*Lett.*) **396**, L1.
11. Hancock, S. *et al.* 1994, *Nature* **367**, 333.
12. For a comprehensive review see White, M., Scott, D. & Silk, J. 1994, *Annu. Rev. Astron. Astrophys.* **32**, 319.
13. Hoyle, F. & Tayler, R. J. 1964, *Nature* **203**, 1108. Peebles, P. J. E. 1966, *Astrophys. J.* **146**, 542. Wagoner, R. V., Fowler, W. A. & Hoyle, F. 1967, *Astrophys. J.* **148**, 3.
14. Yang, J. *et al.* 1984, *Astrophys. J.* **281**, 493. Walker, T. P. *et al.* 1991, *Astrophys. J.* **376**, 51.
15. Burbidge, G. R., Burbidge, E. M., Fowler, W. A. & Hoyle, F. 1957, *Rev. Mod. Phys.* **29**, 547.
16. For a recent review, see Edmunds, M. G. & Terlevich, R. J. (eds.) 1992, *Elements and the Cosmos* (Cambridge University Press).
17. Reeves, H. 1994, *Rev. Mod. Phys.* **66**, 193.
18. Zel'dovich, Y. B. 1982, *Highlights of Astronomy* **6**, 29.
19. Binney, J. & Tremaine, S. 1987, *Galactic Dynamics* (Princeton University Press).

20. Rees, M. J. & Ostriker, J. P. 1977, *Mon. Not. Roy. Astron. Soc.* **179**, 541. Silk, J. I. 1977, *Astrophys. J.* **211**, 638.
21. Lake, G. & Feinswog, L. 1989, *Astron. J.* **98**, 166. Begemann, K. G. 1989, *Astron. Astrophys.* **223**, 47.
22. Ostriker, J. P., Peebles, P. J. E. & Yahil, A. 1974, *Astrophys. J.* (*Lett.*) **193**, L1. Einasto, J., Kaasik, A. & Saar, E. 1974, *Nature* **250**, 309.
23. For a recent review see Ashman, K. M. 1992, *Proc. Astron. Soc. Pacific* **104**, 1109, or Tremaine, S. 1992, *Physics Today* **45**, 28.
24. Kahn, F. D. & Woltjer, L. 1959, *Astrophys. J.* **130**, 705. Einasto, J. & Lynden-Bell, D. 1982, *Mon. Not. Roy. Astron. Soc.* **199**, 67.
25. The first claim for dark matter in clusters was Zwicky, F. 1933, *Helv. Phys. Acta* **6**, 110. For a recent review see Fabian, A. C. (ed.) 1992, *Clusters and Superclusters of Galaxies* (Kluwer).
26. Lynds, R. & Petrosian, V. 1986, *Bull. Amer. Astron. Soc.* **18**, 1014. Soucail, G. *et al.* 1987, *Astron. Astrophys.* **172**, L14.
27. Fort, B. & Mellier, Y. 1994, *Astron. Astrophys. Rev.* **5**, 239 (and references cited therein). Tyson, J. A., Valdes, F. & Werk, R. A. 1990, *Astrophys. J.* (*Lett.*) **349**, L1. Ellis, R. S., Smail, I. & Knieb, J.-P. 1994, *Astrophys. J.* (in press).
28. For a comprehensive recent review of baryonic dark matter, see Carr, B. J. 1994, *Annu. Rev. Astron. Astrophys.* **32**, 531.
29. Paczynski, B. 1986, *Astrophys. J.* **304**, 1.
30. Alcock, C. *et. al.* 1993, *Nature* **365**, 621. Aubourg, E. *et. al.* 1993, *Nature* **365**, 623.
31. Carr, B. J., Bond, J. R. & Arnett, W. D. 1984, *Astrophys. J.* **277**, 445.
32. Refsdal, S. 1964, *Mon. Not. Roy. Astron. Soc.* **128**, 295. Refsdal, S. 1971, *Astrophys. J.* **159**, 357. Chang, K. & Refsdal, S. 1979, *Nature* **282**, 561.
33. Press, W. H. & Gunn, J. E. 1973, *Astrophys. J.* **185**, 397. Blandford, R. D. & Jaroszynski, M. 1981, *Astrophys. J.* **246**, 1. For a comprehensive recent review see Refsdal, S. & Surdej, J. 1994, *Rep. Prog. Phys.* **56**, 117.
34. Applegate, J. & Hogan, C. J. 1985, *Phys. Rev.* **D31**, 3037.
35. Iso, K., Kodama, H. & Sato, K. 1986, *Phys. Lett.* **169B**, 337. Thomas, D. *et. al.* 1994, *Astrophys. J.* **430**, 291 (and references cited therein).

36. Cowsik, R. & McLelland, J. 1973, *Astrophys. J.* **180**, 7.
37. Marx, G. & Szalay, A. 1972, Proc. *Neutrino 72* (Technoinform, Budapest), p. 191.
38. Lyubimov, V. A. *et. al.* 1980, *Phys. Lett.* **394**, 266.
39. For reviews of non-baryonic matter see Kolb, E. W. & Turner, M. S. 1990, *The Early Universe* (Addison Wesley); Ellis, J. R. 1990, in *Physics of the Early Universe*, ed. Peacock, J. A. *et. al.* (Scottish Universities Summer School Publications); or Galcotti, P. & Schramm, D. N. (eds.) 1990, *Dark Matter in the Universe* (Kluwer).
40. For a comprehensive survey of 'correlation functions' see Peebles, P. J. E. 1980, *Large-Scale Structure of the Universe* (Princeton University Press).
41. Bond, J. R., Efstathiou, G. P. & Silk, J. I. 1980, *Phys. Rev. Lett.* **45**, 1980.
42. Peebles, P. J. E. 1982, *Astrophys. J.* **258**, 415. Peebles, P. J. E. 1984, *Astrophys. J.* (*Lett.*) **189**, L51.
43. Harrison, E. R. 1970, *Phys. Rev.* **D1**, 2726. Zel'dovich, Y. B. 1972, *Mon. Nat. Roy. Astron. Soc.* **160**, 1P.
44. Blumenthal, G., Faber, S. M., Primack, J. R. & Rees, M. J. 1984, *Nature*, **311**, 517.
45. Ostriker, J. P. 1993, *Annu. Rev. Astron. Astrophys.* **31**, 689.
46. Frenk, C. *et. al.* 1990, *Astrophys. J.* **351**, 10.
47. Gott, J. R., Gunn, J. E., Schramm, D. N. & Tinsley, B. 1974, *Astrophys. J.* **194**, 543.
48. Shane, C. D. & Wirtanen, C. A. 1967, *Publ. Lick. Obs.* **22**, pt. 1.
49. Maddox, S. J. *et. al.* 1990, *Nature* **349**, 32.
50. Geller, M. J. & Huchra, J. P. 1989, *Science* **246**, 897.
51. Silk, J. I. 1974, *Astrophys. J.* **193**, 525.
52. Lynden-Bell, D. *et. al.* 1988, *Astrophys. J.* **326**, 19.
53. Rowan-Robinson, M. *et. al.* 1990, *Mon. Not. Roy. Astron. Soc.* **247**, 1.
54. For reviews see Bouchet, F. R. & Lachièze-Rey, M. (eds.) 1993, *Cosmic Velocity Fields* (Editions Frontières).
55. Bertschinger, E. & Dekel, A. 1989, *Astrophys. J.* (*Lett.*) **336**, L5. Dekel, A. *et. al.* 1993, *Astrophys. J.* **412**, 1. For a recent review see Dekel, A. 1994, *Annu. Rev. Astron. Astrophys.* **32**, 371.
56. Dekel, A. & Rees, M. J. 1994, *Astrophys. J.* (*Lett.*) **422**, L1.

57. Kaiser, N. 1992, *Astrophys. J.* **388**, 272. Mould, J. *et. al.* 1994, *Mon. Not. Roy. Astron. Soc.* **271**, 31.
58. White, S. D. M. & Frenk, C. S. 1991, *Astrophys. J.* **379**, 52. Lacey, C. & Cole, S. 1993, *Mon. Not. Roy. Astron. Soc.* **262**, 627. Kauffmann, G., White, S. D. M. & Guiderdoni, B. 1993, *Mon. Not. Roy. Astron. Soc.* **264**, 201. Kauffmann, G. & White, S. D. M. 1993, *Mon. Not. Roy. Astron. Soc.* **261**, 921.
59. Pennypacker, C. *et. al.* 1993, *IAU Circ. No.* 5652.
60. Kellerman, K. I. 1993, *Nature* **361**, 134.
61. Turner, E. L. 1990, *Astrophys. J.* (*Lett.*) **365**, L45. Fukugita, M. & Turner, E. L. 1991, *Mon. Not. Roy. Astron. Soc.* **253**, 99.
62. Sandage, A. & Tammann, G. 1990, *Astrophys. J.* **365**, 1. Jacoby, G. H. *et. al.* 1992, *Proc. Astron. Soc. Pacific* **104**, 599. Van den Bergh, S. 1992, *Proc. Astron. Soc. Pacific* **104**, 861.
63. Efstathiou, G. 1990, in *Physics of the Early Universe*, ed. Peacock, J. A. *et. al.* (Scottish Universities Summer School Publications).
64. Bardeen, J. *et. al.* 1986, *Astrophys. J.* **304**, 15.
65. Dekel, A. & Rees, M. J. 1987, *Nature*, **326**, 455.
66. Blanchard, A., Buchert, T. & Klaffl, R. 1993, *Astron. Astrophys.* **267**, 1.
67. Davis, M. *et. al.* 1992, *Nature* **356**, 489.
68. Efstathiou, G., Bond, J. R. & White, S. D. M. 1992, *Mon. Not. Roy. Astron. Soc.* **258**, 1P.
69. Kofman, L., Gnedin, N. Y. & Bahcall, N. A. 1993, *Astrophys. J.* **413**, 1.
70. Klypin, A., Holtzman, J., Primack, J. R. & Regos, E. 1993, *Astrophys. J.* **416**, 1. Haehnelt, M. 1993, *Mon. Not. Roy. Astron. Soc.* **265**, 727.
71. Edge, A. C. *et. al.* 1990, *Mon. Not. Roy. Astron. Soc.* **245**, 559. Henry, J. P. 1992, in *Clusters and Superclusters of Galaxies*, ed. Fabian, A. C. (Kluwer).
72. Hewitt, A. & Burbidge, G. R. 1993, *Astrophys. J. Supp.* **87**, 451.
73. Boyle, B. J., Shanks, T., Fong, R. & Peterson, B. A. 1990, *Mon. Not. Roy. Astron. Soc.* **243**, 1.
74. Hewett, P. C. & Foltz, C. B. 1994, *Proc. Astron. Soc. Pacific* **108**, 113.
75. Fall, S. M. & Efstathiou, G. 1980, *Mon. Not. Roy. Astron. Soc.* **193**, 189.

76. Schmidt, M. 1989, *Highlights of Astronomy* **8**, 31.
77. Soltan, A. 1982, *Mon. Not. Roy. Astron. Soc.* **200**, 115. Phinney, E. S. 1983, Ph.D. Thesis, University of Cambridge.
78. Zel'dovich, Y. B. & Novikov, I. D. 1964, *Dukl. Acad. Nauk SSR* **158**, 811.
79. Salpeter, E. E. 1964, *Astrophys. J. Letts.* **140**, 796.
80. Sargent, W. L. W. *et. al.* 1978, *Astrophys. J.* **221**, 731. Young, P. *et al.* 1978, *Astrophys. J.* **221**, 721.
81. Light, E. S., Danielson, R. E. & Schwarzschild, M. 1974, *Astrophys. J.* **194**, 257.
82. Kormendy, J. 1988, *Astrophys. J.* **335**, 40. Dressler, A. & Richstone, D. O. 1988, *Astrophys. J.* **324**, 701.
83. Tonry, J. L. 1987, *Astrophys. J.* **322**, 632.
84. Jarvis, B. J. & Dubath, P. 1988, *Astron. Astrophys.* **201**, L33.
85. Kormendy, J. *et. al.* 1994, *Astrophys. J.* (in press).
86. Lauer, T. *et. al.* 1994, *Astrophys. J.* (in press).
87. Kormendy, J. 1992, in *Testing the AGN Paradigm*, ed. Holt, S. *et. al.* (American Institute of Physics, New York) p. 23.
88. Rees, M. J. 1988, *Nature* **333**, 523.
89. Kochanek, C. S. 1994, *Astrophys. J.* **422**, 508.
90. Genzel, R. & Townes, C. H. 1987, *Annu. Rev. Astron. Astrophys.* **25**, 377.
91. Genzel, R., Hollenbach, D. & Townes, C. H. 1994, *Rep. Prog. Phys.* **57**, 417.
92. Haehnelt, M. 1994, *Mon. Not. Roy. Astron. Soc.* **269**, 199.
93. Redmount, I. & Rees, M. J. 1989, *Comm. Astrophys.* **14**, 165 and references cited therein.
94. Haehnelt, M. & Rees, M. J. *Mon. Not. Roy. Astron. Soc.* **263**, 168.
95. Navarro, J. & White, S. D. M. 1994, *Mon. Not. Roy. Astron. Soc.* (in press).
96. Hubble, E. 1936, *The Realm of the Nebulae* (Yale University Press).
97. Chandrasekhar, S. 1975, Lecture reprinted in *Truth and Beauty* (Chicago University Press 1987), p. 54.
98. Gunn, J. E. & Peterson, B. A. 1965, *Astrophys. J.* **142**, 1633.
99. Rees, M. J. 1986, *Mon Not. Roy. Astron. Soc.* **218**, 25P. Ikeuchi, S. 1986, *Astrophys. & Sp. Sci.* **118**, 509.
100. Miralda-Escude, J. & Rees, M. J. 1993, *Mon. Not. Roy. Astron.*

Soc. **260**, 617. Miralda-Escude, J. & Rees, M. J. 1994, *Mon. Not. Roy. Astron. Soc.* **266**, 343.

101. Smette, A. *et. al.* 1992, *Astrophys. J.* **389**, 39.
102. Mo, H. J., Miralda-Escude, J. & Rees, M. J. 1993, *Mon. Not. Roy. Astron. Soc.* **264**, 705.
103. Rephaeli, Y. & Szalay, A. 1981, *Phys. Lett. B* **106**, 73. Scott, D., Rees, M. J. & Sciama, D. W. 1991, *Astron. Astrophys.* **250**, 295.
104. Tegmark, M. & Silk, J. I. 1994, *Astron. Astrophys.* **423**, 529.
105. Scott, D. & Rees, M. J. 1990, *Mon. Not. Roy. Astron. Soc.* **247**, 510.
106. Swarup, G. 1992, *Giant Metre Wave Telescope* (TIFR Publications).
107. Kronberg, P. P. 1994, *Rep. Prog. Phys.* **57**, 325.
108. Zel'dovich, Y. B., Ruzmaikin, A. A. & Sokolov, D. D. 1983, *Magnetic Fields in Astrophysics* (Gordon & Breach).
109. Mestel, L. 1994, in *Cosmical Magnetism*, ed. Lynden-Bell, D. (Kluwer).
110. Ratra, B. 1992, *Astrophys. J.* (*Lett.*) **391**, L1.
111. Mestel, L. & Spitzer, L. 1956, *Mon. Nat. Roy. Astron. Soc.* **116**, 503.
112. Biermann, L. 1950, *Z. Nat.* **5a**, 65.
113. Kennel, F. & Coroniti, F. V. 1984, *Astrophys. J.* **283**, 694.
114. Chambers, K. C., Miley, G. K. & van Breugel, W. J. M. 1990, *Astrophys. J.* **363**, 32.
115. Parker, E. N. 1979, *Cosmic Magnetic Fields* (Oxford University Press).
116. Kibble, T. W. B. 1976, *J. Phys.* **A9**, 1387.
117. For a comprehensive review see Valenkin, A. & Shellard, P. 1994, *Cosmic Strings and other Topological Defects* (Cambridge University Press).
118. Kaiser, N. & Stebbins, A. 1984, *Nature* **310**, 391. Coulson, D. *et. al.* 1994, *Nature* **368**, 27.
119. Hogan, C. J. & Rees, M. J. 1984, *Nature* **311**, 109.
120. Taylor, J. H. 1992, *Phil. Trans. Roy. Soc.* **A341**, 117.
121. Peebles, P. J. E. 1987, *Nature* **327**, 210.
122. Dicke, R. H. & Peebles, P. J. E. 1979, in *General Relativity: An Einstein Centenary Survey*, ed. Hawking, S. W. & Israel, W. (Cambridge University Press).

123. Guth, A. 1981, *Phys. Rev.* **D23**, 347.
124. Narlikar, J. V. & Padmanabhan, T. 1991, *Annu. Rev. Astron. Astrophys.* **29**, 325.
125. Liddle, A. R. & Lyth, D. H. 1992, *Phys. Rep.* **231**, 1.
126. Adams, F. C. *et. al.* 1992, *Phys. Rev.* **D47**, 426.
127. Starobinski, A. A. 1985, *Sov. Astron. Lett.* **11**, 113. Crittenden, R. *et. al.* 1993, *Phys. Rev. Lett.* **71**, 324.
128. Linde, A. D. 1990, *Particle Physics and Inflationary Cosmology* (Harwood, Switzerland).

Some further reading

Texts

Kolb, E. W. & Turner, M. S. 1990, *The Early Universe* (Chicago University Press).

Padmanabhan, T. 1993, *Structure Formation in the Universe* (Cambridge University Press).

Peebles, P. J. E. 1993, *Principles of Physical Cosmology* (Princeton University Press).

Proceedings of conferences or schools

Crampton, D. (ed.) 1991, *The Space Distribution of Quasars* (Astronomical Society of the Pacific, Conference Series. No. 21).

Fabian, A. C. (ed.) 1992, *Clusters and Superclusters of Galaxies* (Cambridge University Press).

Hawking, S. W. & Israel, W. (eds.) 1987, 300 *Years of Gravitation* (Cambridge University Press).

Nobel Symposium 79 on 'Birth and Early Evolution of our Universe', *Physica Scripta* **T36**, 1991.

Peacock, J. A., Heavens, A. F. & Davies, A. T. (eds.) 1989, *Physics of the Early Universe* (Hilger).

Prantzos, N., Flam, E. & Casse, M. (eds.) 1994, *The Origin and Evolution of the Elements* (Cambridge University Press).

Rocca-Volmerange, B. *et. al.* (eds.) 1993, *First Light in the Universe: Stars or QSOs* (Editions Frontières).

Rubin, V. C. & Coyne, G. V. (eds.) 1988, *Large-Scale Motions in the Universe* (Princeton University Press).

Schramm, D. N. (ed.) 1993, Proceedings of the National Academy of Sciences Conference on Cosmology, *Proc. Nat. Acad. Sci.* Vol. 90.

Shanks, T. *et. al.* (eds.) 1991, *Observational Tests of Cosmological Inflation* (Kluwer).
Thronston, H. A. & Shull, J. M. (eds.) 1993, *The Evolution of Galaxies and their Environments* (Kluwer).

Author index

Note The superscript numbers refer to the corresponding reference numbers on pp. 125–31.

Subject index